P9-DTT-471

OPERATION
MERCURY

0 11557 03506 3

The Stackpole Military History Series

THE AMERICAN CIVIL WAR
Cavalry Raids of the Civil War
Ghost, Thunderbolt, and Wizard
Pickett's Charge
Witness to Gettysburg

WORLD WAR II
Armor Battles of the Waffen-SS, 1943–45
Army of the West
Australian Commandos
The B-24 in China
Backwater War
The Battle of Sicily
Beyond the Beachhead
The Brandenburger Commandos
The Brigade
Bringing the Thunder
Coast Watching in World War II
Colossal Cracks
A Dangerous Assignment
D-Day to Berlin
Dive Bomber!
A Drop Too Many
Eagles of the Third Reich
Exit Rommel
Fist from the Sky
Flying American Combat Aircraft of World War II
Forging the Thunderbolt
Fortress France
The German Defeat in the East, 1944–45
German Order of Battle, Vol. 1
German Order of Battle, Vol. 2
German Order of Battle, Vol. 3
The Germans in Normandy
Germany's Panzer Arm in World War II
GI Ingenuity
The Great Ships
Grenadiers
Infantry Aces
Iron Arm
Iron Knights
Kampfgruppe Peiper at the Battle of the Bulge
Kursk
Luftwaffe Aces
Massacre at Tobruk

Mechanized Juggernaut or Military Anachronism?
Messerschmitts over Sicily
Michael Wittmann, Vol. 1
Michael Wittmann, Vol. 2
Mountain Warriors
The Nazi Rocketeers
On the Canal
Operation Mercury
Packs On!
Panzer Aces
Panzer Aces II
Panzer Commanders of the Western Front
The Panzer Legions
Panzers in Winter
The Path to Blitzkrieg
Retreat to the Reich
Rommel's Desert Commanders
Rommel's Desert War
The Savage Sky
A Soldier in the Cockpit
Soviet Blitzkrieg
Stalin's Keys to Victory
Surviving Bataan and Beyond
T-34 in Action
Tigers in the Mud
The 12th SS, Vol. 1
The 12th SS, Vol. 2
The War against Rommel's Supply Lines
War in the Aegean

THE COLD WAR / VIETNAM
Cyclops in the Jungle
Flying American Combat Aircraft: The Cold War
Here There Are Tigers
Land with No Sun
Street without Joy
Through the Valley

WARS OF THE MIDDLE EAST
Never-Ending Conflict

GENERAL MILITARY HISTORY
Carriers in Combat
Desert Battles
Guerrilla Warfare

OPERATION MERCURY

The Battle for Crete, 1941

John Sadler

STACKPOLE
BOOKS

Published in paperback in 2008 by
STACKPOLE BOOKS
5067 Ritter Road
Mechanicsburg, PA 17055
www.stackpolebooks.com

OPERATION MERCURY: THE BATTLE FOR CRETE, 1941 by John Sadler
was originally published in hardcover by Pen & Sword Military, an imprint
of Pen & Sword Ltd., South Yorkshire, England. Copyright © 2007 by
John Sadler. Paperback edition by arrangement with Pen & Sword Ltd.

Cover design by Tracy Patterson

Printed in the United States of America

10 9 8 7 6 5 4 3 2 1

ISBN 0-8117-3506-0 (Stackpole paperback)
ISBN 978-0-8117-3506-3 (Stackpole paperback)

The Library of Congress cataloging-in-publication data:

Sadler, John, 1953–
  Operation Mercury : the battle for Crete, 1941 / John Sadler.
      p. cm.
  Originally published: Barnsley, South Yorkshire : Pen & Sword Military, 2007.
  Includes bibliographical references and index.
  ISBN-13: 978-0-8117-3506-3 (pbk.)
  ISBN-10: 0-8117-3506-0 (pbk.)
  1. Operation Mercury, 1941. I. Title.
  D764.3.O64S23 2008
  940.53'1844—dc22

                                                                    2008015291

Dedication

This book is dedicated to the memory of Alexander Bruce Cowper, 2077985 Royal Engineers, killed in action 28 May 1941 during the Battle for Crete and whilst serving with Layforce. Like so many others he has no known grave.

> The young hate the old
> Yet stumble after them.
> Stand for New Zealand!
> Yelled Kippenberger,
> Country lawyer turned Brigade Commander
> And conveyed ten thousand miles
> To practise heroics
> On a deadly Cretan hillside.
> Stand for New Zealand!...
> Over the fraudulent field of death
> Forward for New Zealand!
> You random assembly of farm labourers,
> Clerks, rouseabouts, shearers, barmen,
> Salesmen, commercial travellers, storemen,
> Mechanics, musterers, drivers, factory hands,
> And seasonal workers pressed into temporary khaki...
> We are infantry of mettle
> Reputed for steadfastness in the attack,
> The highest produce of a country that
> Breeds men with the animal virtue of blind courage
> In the willing service of the herd.
> Our lifestyle and instincts instruct us
> More cogently than any military precepts.
> Forward for New Zealand!

Quoted by Tony Simpson in *Operation Mercury* pp254-5.

Contents

Acknowledgements

The author wishes to acknowledge the assistance of the Durham Light Infantry Museum in Durham; The County Durham Records Office; The North East Military Vehicle Museum; the members of The Northern World War Two Association; the Imperial War Museum; the staff of the Naval Museum in Chania and George Andreas Hatzidakis of the War Museum, Askifou; Steve Shannon at the DLI Museum; David Fletcher of the Tank Museum Bovington; Roberta Twinn of the Discovery Museum Tyneside; Sara Bevan of the Imperial War Museum; Rod Mackenzie of the Argyll and Sutherland Highlanders; Thomas B. Smyth of the Black Watch; Paul Evans of the Royal Artillery Museum and Anna Tiaki of the Alexander Turnbull Library, New Zealand. Thanks are also due to my sister-in-law Mary Towns for providing the 'family' connection; to the late Nigel Porter; Timothy Norton; Kit Pumphrey, Sir Lawrence Pumphrey and Fred Mason for advice and assistance regarding the role of the Northumberland Hussars and for providing access to unpublished papers and, as always, to my wife, Ruth, who bravely endured a rather hair-raising drive down the spectacular switchbacks of the Imbros Gorge. All errors are, of course, entirely the responsibility of the author.

John Sadler.

Northumberland, 2006.

Let me forget – let me forget,
I am weary of remembrance,
And my brow is ever wet,
With the tears of my remembrance,
With the tears and bloody sweat,
Let me forget.

If ye forget – if ye forget,
Then your children must remember,
And their brow be ever wet,
With the tears of their remembrance,
With the tears and bloody sweat.

If ye Forget by G.A. Studdert-Kennedy

A Note on Spelling

Students of the Battle for Crete could be excused for a degree of mounting frustration over the common spelling of place names which differs rather widely from account to account. Thus Chania can appear as Hania or Canea, Heraklion as Iraklion, Herakleion or Iraklio and so on. The difficulty lies in the fact that there is no generally recognised system for transcription of modern Greek to English. Greek publications themselves exhibit inconsistencies. I have therefore opted to follow Baedeker and retain the accepted orthography of classical place names.

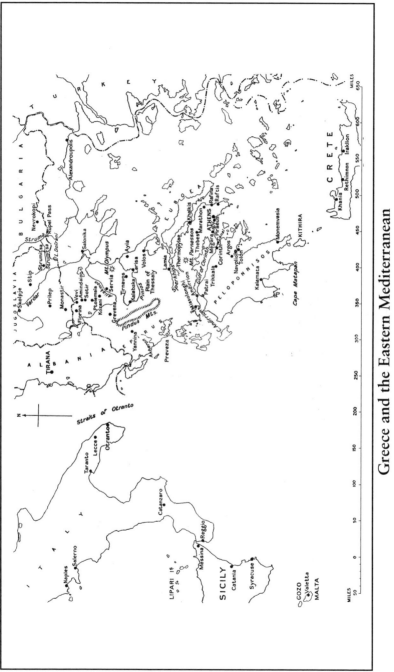

Greece and the Eastern Mediterranean

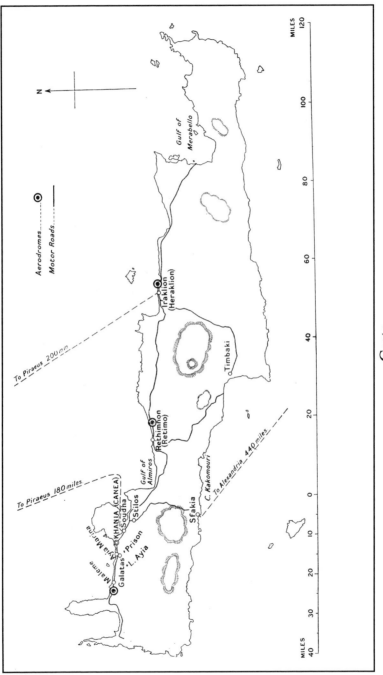

Crete

CHAPTER 1

A Parallel War

When a Campaign ends in disappointment or in disaster, every pundit at once asks why it was ever started. But history often records that a failure is as glorious and, in the long view, as useful as a victory.[1]

Bone weary, part demoralised, hungry and thirsty, the survivors of the Crete garrison surveyed the steep, seemingly unending climb from the stifling heat of the coastal plain to the rim of the Askifou plateau in the final days of May 1941. They could be forgiven if they failed to share the official historian's bent for propaganda. These exhausted and beaten men would be unlikely to have perceived anything vaguely useful in their defeat. Major General F.F.C. Graham of the Commandos painted a substantially less edifying view:

> The road was jammed with troops in no formed bodies shambling along in desperate haste. Dirty, weary and hungry, they were a conglomeration of Australians, a few New Zealanders and British, and Greek refugees. They had only one thing in common and that was a desire to get as far as possible from Canea – a rabble one could call them, nothing else ... desperately we pushed our way on to the road and tried to push past the motley throng which straggled over it. All day the sky was thick with enemy aircraft, in many cases flying at only a few hundred feet and every now and again coming down to bomb and machine gun the troops trudging along the road. All day the stream of the retreat flowed steadily but wearily on. When enemy aircraft approached, the bulk of the men tried to scatter off the road or hid in the ditches; some, impervious to threats such as: 'Lie down you bastard or I'll fucking shoot you,' bore steadily

1

on. [We] lay with the sun on our backs straining our eyes to catch the first sign of enemy movement to the north or in the mountains to the west.[2]

Now there is a well metalled road from Souda through Vrysses leading up to the plateau through a series of hairpin bends; the land stretches away, desolate in summer, sparse grass and scrub, strewn with rocky outcrops and loose scree. The sun is a demon, a few scattered goats cling to the hillside, their bells the only melody in the thyme scented air. There is virtually no water and the ascent is steep and exposed in brilliant light; an air gunner's dream.

The Askifou Plateau is a bare bowl surrounded by barren slopes, leading to the greater heights of the Lefka Ori, the White Mountains, that crown the island's backbone. The stump of an ancient Crusader castle stands astride a fat knoll to the east, silent witness to a long history of occupation.

Once across the plateau the ground disappears into the sudden drop of the Imbros Gorge, a primeval landscape of plunging hillsides. The modern road embarks on the dizzying and seemingly interminable series of hairpins leading down to the narrow coastal plain 2,500 feet below. It looks much further, the scattered habitations like white dots on an ochre, chequered board, the descent laborious.

This is the edge of Europe, the southern rim; south across the hazed blue waters lies Egypt, another continent, 200 miles distant. Sphakia, now a compact jumble of holiday lets and apartments, crowding a narrow harbour bedecked with the froth of mass tourism, clings to the hillside.

It was here that the survivors waited in their thousands, their nightmare route over the pass littered with the detritus of defeat, abandoned rifles, packs, uniform and ruined vehicles. Their only hope lay in the Navy, the pursuing Germans continually and determinedly held off by a dogged rearguard. For many, this was their second such debacle, the first the expeditionary force's flight from mainland Greece in April.

Yet, only a few days previously, General Bernard Freyberg the C.-in-C. had sent a highly confident cable to GHQ Middle East:

Have completed plan for defence of Crete and have just returned from final tour of the defences. I feel greatly encouraged by my visit. Everywhere all ranks are fit and morale is high ... I do not wish to be overconfident but I feel that we will at least give excellent account of ourselves. With help of Royal Navy I trust Crete will be held.[3]

Crete was not held, the island was lost, the second such catastrophe to afflict British forces that disastrous spring of 1941. The fall of Crete was the final act in a two act tragedy of British involvement in the Balkans.

> Greece is what everybody knows, even in absentia, even as a child or as an idiot or as a not-yet-born. It is what you expect the earth to look like given a fair chance. It is the subliminal threshold of innocence. It stands, as it stood from birth, naked and fully revealed. It is not mysterious or impenetrable, not awesome, not defiant, not pretentious. It is made of earth, fire and water. It changes seasonally with harmonious, undulating rhythms. It breathes, it beckons, it answers.
>
> Crete is something else, Crete is a cradle, an instrument, a vibrating test tube in which a volcanic experiment has been performed. Crete can hush the mind, still the bubble of thought.[4]

Crete is the biggest of the Greek islands after Cyprus, the fifth largest in the whole of the Mediterranean Sea. It lies west to east in the Southern Aegean, 160 miles (260 km.) in length but with a narrow girth of between 7.5 and 37 miles (12 to 60 km.). Some 63 miles (100 km.) south-east of the Peloponnese it forms the central link in a chain of islands spanning the watery gap between mainland Greece and Turkey.

Geologically, its position is a perilous one, the shelf of the Aegean plate is a zone of considerable geological tension and this has produced a series of devastating earthquakes which have blighted the island's history, the most recent being a quake which registered 6.1 on the Richter scale in 1994. Happily there was no loss of life or major structural damage but an earlier tremor in 1926 caused widespread devastation.

Such a volatile geological heritage has produced the three main mountain ranges that form the stiff backbone of the island, split in places by dramatic gorges such as appear at Imbros and Samaria. To the west the sweeping heights of the Lefka Ori rise dramatically to a majestic height (8,045 ft./2452 m.). Mount Ida in the centre forms the highest peak at 8,058 ft./2456 m. Further east the Dikti range surrounds the high plain of the Lassithi Plateau – birthplace of Zeus and indeed such an elemental landscape is a fitting home for the ancient Gods.

Above Sitia on the tip of the east coast, the Thriptis hills rise to 4,843 ft./1476 m. The limestone formations are honeycombed with caves, sculpted with stunning stalactite formations. In summer the network of twisting valleys and rivulets run dry; only in winter or during a wet spring do the watercourses fill up. Spring tends to come quite early, certainly by April, and the hot blaze of the southern Mediterranean summer persists usually until late September or October, with scant rainfall between.

The long south coast is a romantic wilderness with the high cliffs sweeping sheer to the shore leaving, at best, only a narrow coastal margin. The northern flank is more populous and here lie the harbours of Chania, Rethymnon and Heraklion – it is here that successive waves of migrants, settlers, refugees and invaders have left their mark.

Crete has a very long and dramatic history; in the Bronze Age the Minoan rulers created a unique palace culture, their unwalled cities inviolate behind the might of their swift oared galleys. Earthquakes and the violent eruption of ancient Thera (Santorini) may have contributed to the collapse of the Minoan civilisation and its absorption in the wider Mycenaean culture. Idomeneus, the prince of Crete, is said to have led a commanded body of elite archers to assist the Greeks at the siege of Troy.

The island languished during the relative dark age before the dawn of the classical era; it later came within the hegemony firstly of Rome and, latterly, Byzantium. In the years 826/827 it fell to the Arabs before being recovered, in 961, by the Eastern Empire. The shock of the rapacious Fourth Crusade in 1204 left the island in the sway of the Venetians and they stayed for some 450 years. Though much resented, their monuments abound and their influence still haunts the cities and ports of the north coast.

Islanders and occupiers made common cause against the swelling might of the Ottoman Turks. To no avail, however, for, by 1669, their conquest was complete and Crete remained firmly under the Muslim yoke until 1898. Once the mainland of Greece obtained its independence, after a long and savage struggle, the Cretans frequently rebelled. These were murderous affairs of guerrilla ambushes and merciless reprisal, the forge in which the fierce independence and uncompromising honour of the mountain men was hardened.

The Cretans were no amateurs at resistance, nor were they a

civilian population to be cowed by the conqueror's military might –
a phenomenon that would astound and enrage the Germans, used to
an effortless dominance over defeated peoples. Loyal to family and
kin, imbued with the tribal virtues of hardiness, hospitality to
strangers and general contempt for centralised authority, deeply
devout and fiercely independent, adept at the pernicious ritual of the
blood feud, they made remarkable allies and implacable foes.

These Palikari cut a dramatic dash, clad in the mountain
shepherd's uniform of black, knee length leather jackboots, baggy
woollen breeches, embroidered dark waistcoat, full sleeved shirts,
wound headscarves and a long bladed, bone handled yataghan
dagger thrust into a coloured sash.

> Crete is in truth a continent on its own and should be seen as
> such by visitors
> Crete, the oldest Europe, has its youngest people, bursting with
> vitality and originality, as insatiable in its thirst for life as it is
> unhesitating in squandering itself. So many sediments of history
> weight down on Crete and yet it shows no weariness, no trace of
> exhaustion: an air of wisdom – such wisdom that it can commit
> every kind of folly.[5]

In 1898 an alliance of the European powers forced Turkey to cede a
measure of independence, though retaining sovereignty, with Prince
George of Greece being appointed as a form of high commissioner.
The islanders liked neither prince nor compromise and, eight years
later, the patriot politician Eleftherios Kyriakis Venizelos led a
popular uprising whose objective was to force a full union with
Greece. Although he became national premier in 1910, full integra-
tion was delayed until 30 May 1913.

Venizelos, virulently anti Turk, led his country in the successful
prosecution of the Balkan Wars of 1912-1913. During the Great War
he leaned toward the Allied cause, mainly in the hope of securing a
greater slice of the Ottoman carcass. Conversely, King Constantine I
sought, unsuccessfully, to retain full neutrality, his sympathies lying
more toward Germany.

The British were not impressed and casually occupied firstly
Lemnos as a base for Gallipoli and, latterly with France, the great,
malarial sprawl of the encampment at Salonika, from where, ostensibly, it was proposed to assist the Serbs against Austria.

Temporarily ousted, Venizelos riposted with typical vigour and forced the King's abdication, thereafter placing Greece firmly in the Allied camp.

Earlier gains were squandered in the disastrous wars of 1920-1923; the Young Turks responded vigorously to fresh aggression and Greece lost not only her outposts in Asia Minor but also the islands of Imbros and Tenedos. A great rush of refugees, driven out by the Turks, flooded into Crete whilst the indigenous Muslims fled, abandoning their final toehold and leaving only echoes of their architecture in shuttered balconies and the odd crumbling minaret.

Greece became a republic in 1924, a state which appealed to most Cretans, distrustful of monarchies and rather contemptuous of their mainland cousins. Venizelos led a liberal administration until the uncertainties of the economic slump produced a swing to the right. The monarchy was restored and, after a final term in office, the ageing lion was ousted by the hardline General Metaxas, dying in exile in Paris in 1936.

The new premier, whilst an ardent nationalist, saw no need to retain the democratically elected assembly, preferring the sinister autocracy of the police state, with the King as a compliant figurehead. This repressive regime was stoutly resisted by the independently minded Cretans, to whom the paramilitary gendarmerie became a symbol of oppression. Attempts were made to disarm the islanders and the lingering suspicion would bear bitter fruit when Cretans were starved of arms by a regime that feared to arm its own people.

In 1940 the island was still something of a backwater, good roads were virtually non-existent, only the single ribbon of the coastal highway linked the towns along the north coast. The economy remained substantially agrarian, peasant farmers working the olive groves that covered 41 per cent of the cultivable acres, and producing 120,000 tons of oil per annum. The remainder of the useful land was given over to grazing for sheep, goats, pigs and poultry.

Politically, Crete was divided into the four administrative sectors ('nomos'/'nomi') of Chania, Rethymnon, Heraklion and Lassithi (Ayios Nikolaos). Chania, the second largest of the towns after Heraklion, and birthplace of Venizelos, was the traditional, Venetian capital. The superb harbour was surrounded by massive artillery walls (which largely survive), the delightful winding alleys of the old town an eclectic blend of Italianate and Turkish design.

Both Heraklion, the largest settlement, and Rethymnon exhibited similar provenance, a riot of ancient streets within the span of the Venetian walls.[6] Until the dramatic events of 1941 began to unfold, Crete was regarded as militarily insignificant. Its main strategic attraction, particularly from the British view at GHQ Middle East in Cairo, was the magnificent anchorage of Souda Bay, just east of Chania. Possession of so extensive and fine a harbour could influence the balance of naval supremacy in the Eastern Mediterranean.

It could be argued that the fall of France and the evacuation at Dunkirk, apparently catastrophic, did, in fact, confer an element of strategic advantage upon Britain. Freed from the dire attrition of obligations to continental allies, such as had enmeshed Imperial forces during the Great War with the meat grinder of the Western Front, Britain could fall back upon her traditional strengths. These were an all powerful navy and a resolute air force that had, in the summer of 1940, successfully defied the Luftwaffe's best efforts.

Germany's failure to crush Britain presented the Nazi high command with a limited range of strategic options. To all intents and purposes the war in the west was won, as Hitler intimated when writing to Mussolini.[7] All that remained was the final push against a moribund and defeated England. Similar assurances were offered to the Russians who appear to have been more sceptical. The key question for Berlin was whether they could afford to turn the swollen and victorious Wehrmacht east to settle with Russia while the British Empire, technically at least, was still in the ring.

War on two fronts was the strategic nightmare which had haunted German military planners since the end of the nineteenth century. The opening moves of the Great War, in the late summer of 1914, had been carefully choreographed to achieve a swift victory in the west before the Kaiser's legions had to turn and face the slower but monolithic threat from Russia. Hitler was in no doubt that, to achieve the eastern border of his greater Reich, he would have to deal with Stalin and the Red army; it was merely a question of when.

Hitler, therefore, had to decide if he could afford to leave an impotent but undefeated Britain in his rear or should he bide his time, consolidate his resources, particularly in the air, so he could finally crush the RAF and make the prospect of invasion real enough to force Churchill to negotiate a peace.

Nor was east the only direction which counted; Hitler had, yet, hopefully through diplomacy, to 'sell' the concept of greater Germany to the world at large and particularly to America. He was obviously aware that Britain's hope of any renewed offensive capacity lay with the United States and Russia, the unlikeliest of bed-fellows.

Even before the aerial defeat of the Luftwaffe in the summer skies, in July 1940, Hitler had laid out his reasoning to his generals:

> In the event that invasion does not take place, our efforts must be directed to the elimination of all factors that let England hope for a change in the situation – Britain's hope lies in Russia and the United States. If Russia drops out of the picture, America too is lost for Britain, because the elimination of Russia would greatly increase Japan's power in the Far East. Decision: Russia's destruction must therefore be made a part of the struggle – the sooner Russia is crushed the better'.[8]

Admiral Raeder, who headed the Kriegsmarine, put forward another option which was to attack Britain and bring her finally to her knees, not through an assault on England but by striking at her empire. The string of defeats in Norway and France had convinced the Axis powers that the British Empire was a rotten hulk, a bankrupt corporation, just waiting for the receivers to move in:

> The British have always considered the Mediterranean the pivot of their world empire [Raeder wrote] ... While the air and submarine war is being fought out between Germany and Britain, Italy, surrounded by British power, is fast becoming the main target of attack. Britain always attempts to strangle the weaker enemy. The Italians have not yet realised the danger when they refuse our help ... The Mediterranean question must be cleared up during the winter months ... The seizure of Gibraltar ...The dispatch of German forces to Dakar and the Canary Islands ... Close co-operation with Vichy ... Having secured her western flank by these measures Germany would support the Italians in a campaign to capture the Suez Canal and advance through Palestine and Syria. If we reach that point Turkey will be in our power. The Russian problem will then appear in a different light. Fundamentally Russia is afraid of Germany. It is doubtful whether an advance against Russia in the north will then be necessary.[9]

With the collapse of France the strategic position in the Mediterranean had changed significantly. It was French naval power that would have secured the western end and the position of the colonies in North Africa and, particularly, Syria was now critical. Would Vichy remain neutral or align with the conqueror thus exerting further pressure on the overstretched British?

The Tripartite Alliance which linked Japan's far eastern power to the Axis and which was agreed in September 1940, would, in Ribbentrop's confident assertion, keep America firmly focused on the Pacific and deter any overt intervention in Europe. With Raeder's concept in mind Hitler made overtures both to Franco in Spain and to Petain.

After a tortuous nine hour meeting with the Spanish Generalissimo at Hendaye, the Führer declared that he'd rather have several teeth removed than repeat the experience. Franco was lukewarm and expected any commitment to be rewarded by the distribution of French colonies in North Africa. This was not something calculated to appeal to Vichy.

If it was made known that their participation in a combined attack on British interests would result in a hiving off of territory to fascist Spain, then where was their incentive? Hitler, showing no great enthusiasm himself, could only suggest that the French could be compensated by the grant of captured British provinces. Franco, cannily, remained sceptical over Axis prospects of finally defeating Britain.

There was also the question of the oil which in war, as Clemenceau pronounced, 'is as necessary as blood'.[10] Without an adequate and continuous supply of crude oil no state could meet the huge demands of modern warfare. In 1939 Britain was importing some 9 million tons of crude, the bulk of which flowed from Iran, Iraq and the USA. By early 1941 oil reserves had fallen to alarmingly low levels; a failure of supply would force Britain to seek terms as surely as a renewed and successful air offensive.

It was estimated, by the Petroleum Board that between spring of 1940 and 1941 some 14 million tons of oil would be required – the USA could be counted on to supply less than half of this and the oil producers were not swayed by Anglophile considerations – they wanted cash on the barrel; precious dollars that Britain could ill afford to disburse.

The Shah of Iran had seized on the urgency to extort fresh concessions; supplies from the USA were subject to sickening wastage through U-Boat attacks. As the Vichy administration in Syria stood adjacent to the vital oilfields in the Middle East and the vulnerable, key refinery at Haifa, their sympathies were of some concern. In a memorandum of 3 August 1940 the Colonial Office provided a neat précis of the position:

> Our first aim on the collapse of France was to induce the French colonies to fight on as our allies. We established close contact with the local French administrations and offered substantial financial inducements. The reaction leaves no doubt that there are French elements ready to rally to our side. But the Vichy government is in a position to exercise strong pressure on the local officials who have been in a defeatist and wavering frame of mind; and there is no doubt that the official policy of the local administrations is now one of obedience to Vichy and refusal to co-operate with us.[11]

This was hardly encouraging – if French naval power, based in the North African ports, was to be added to that of the formidable Italian fleet then the balance of naval power in the Mediterranean would be upset. The oil tankers sailing from Iran would be vulnerable and Britain's entire position in the Middle East under serious threat. Already Mussolini was massing troops in Cyrenaica for a thrust into Egypt, their numbers vastly exceeding those of the defenders.

In November 1940 the Fleet Air Arm sallied against Italian capital ships sheltering in Taranto and scored a signal success, a further engagement off Cape Spartivento on 27th of that month reinforced British superiority but the threat from a compliant Vichy remained.

Germany too, had concerns over its own supply of oil, every bit as pressing as England's. Lacking natural resources and the advantages of an established international network, starved of cash reserves by the crippling burden of war debt, she looked eastwards to the Balkans and the rich Romanian oilfields around Ploesti. By clever and ruthless economic manipulation Germany had succeeded in exerting a measure of fiscal control over her Balkan suppliers.

This dependency was not lost on Britain whose own experts had forecast a crisis of supply. At the outbreak of hostilities Germans had reserves totalling some 3 million tons of crude, yet demand over each of the following three years was predicted to nearer 10 million tons. To avoid shortfalls the Germans were obliged to increase imports from Russia and to make use of captured stocks.

The possibility of a British attack on the Ploesti oilfields was a tactical nightmare which loomed large in the minds of German planners – such an offensive had been considered but the practical difficulties were considerable. To facilitate aerial bombardment the British would need to utilise forward bases located on either Greek or Turkish soil.

It is highly probable that Hitler, at this stage, was not seeking any direct military involvement in the Balkans. The area had no immediate strategic importance beyond the need to secure Germany's oil supply and the Führer was reluctant to become embroiled in the morass of Balkan politics at a time when his attention was fixed on the east.

Added to the traditional swirl of antipathies were the fresh sores opened by the re-drafting of the maps, undertaken by the victorious allies in 1918, dividing the carcass of the old Habsburg Empire. Yugoslavia was a composite state comprising a mix of races and faiths and dominated by the Orthodox Serbs; Hungary smarted from the loss of Transylvania, ceded to Romania from whom the Soviets also sought to recover Bessarabia.

Rising tensions between Hungary and Romania so alarmed Hitler that he became personally involved, bullying the squabbling parties into acceptance of German arbitration. The resulting enforced settlement, the Vienna Award, cost the Romanians dear but restored some semblance of calm. The Nazis refused to take chances, however, and from September 1940 their military presence around the vital oilfields, with the acquiescence of a compliant regime, was steadily increased.

No sooner has the hapless Romanians been stripped of territories in Transylvania than the Bulgarians began pestering for the acquisition of Southern Dobrudja. Once again Hitler obliged Bucharest to submit though he did, this time, offer the guarantee that no further limbs would be shorn from Romania's reduced torso.

Such arbitrary re-drawing of Balkan borders naturally antagonised the Russians who were accustomed to reserving this for themselves. The Russian Foreign Minister Molotov was quick to reprove the German ambassador for a perceived violation of their Non Aggression Pact. The very last thing Hitler sought, at this point, was to provide the Soviets with a casus belli – war there would be, but only when Germany was ready and fully deployed.

Aside from possible British aggression, the other fact which alarmed Hitler was the likely intentions of his Italian ally. Il Duce was steadily becoming disaffected with the relegation of his country's role in the war. Where were the great gains he had hoped for, the triumphal marches of fascist armies through captured Allied dependencies? Now very much the junior partner in the Axis Alliance, Mussolini was hungry for spoils.

The relationship between the two states was further strained by a clear element of mistrust; the Germans, or at least elements of the high command, were deeply suspicious of their ally's competence in the intelligence war, to the extent that outright treachery was mooted.[12]

Aware that Italy was looking covetously at Greece, Hitler took pains to ensure his fellow dictator was aware that Germany was steadfastly opposed to any military adventure. In August 1940 von Ribbentrop had warned the Italian ambassador that Greece was not on anyone's agenda. Despite this, suspicions persisted, all too well founded as it turned out, in the face of a string of bland denials.

In all of this the attitude of the Americans was crucial. President Roosevelt was convinced of the evils of fascism and the need to sustain the British war effort. Initially the majority of the electorate was, at best, lukewarm and opposed to any firm commitment. The President was constrained to keep most of his discussions with Churchill out of the public arena, especially as he was obliged to fight an election in the autumn of 1940.

Despite the abundant and growing evidence of Nazi tyranny and wholesale oppression, Germany still enjoyed the support of a vociferous minority in the USA. The aviator Charles Lindbergh acted as the focus for a group styled 'America First' – fiercely isolationist and dismissive of British prospects, it openly advocated an accord with Hitler and the Nazis. Joseph Kennedy and the Irish lobby within Roosevelt's own party were rabid Anglophobes.

As ambassador to St. James, Kennedy was less than a popular success, seeking to gain personal business advantage as a condition precedent to continued aid and openly touting for an armistice. The president was considerably embarrassed and recalled Kennedy. Repellent as his attitude and conduct were, he was not, by any means, in a minority; many US industrialists shared his leanings.

Gerhard Westrick, a German agent with a trade attaché cover, liaised with numerous leading figures from the sphere of business and commerce, offering significant trading opportunities for US businesses with the Greater German Reich. A number of key industrial figures appeared convinced by his blandishments.[13]

By the close of 1940, a year of serial disasters for England, livened only by the twin deliverance of Dunkirk and the Battle of Britain, the war cabinet was desperate for US support, the continuance of which was beyond the beleaguered and depleted capital of the state and its Empire. Churchill penned an eloquent and detailed essay to Roosevelt, stressing the need for unencumbered aid:

> I believe you will agree that it would be wrong in principle and mutually disadvantageous in effect, if at the height of this struggle Great Britain were to be divested of all saleable assets, so that after the victory is won with our blood, civilisation saved, and the time gained for the United States to be armed against all eventualities, we should stand stripped to the bone.[14]

This passionate plea touched a continuing chord in the president and became the inspiration for the subsequent Lease-Lend scheme. More immediately, and within a couple of days, Roosevelt had dispatched his personal representative, Colonel William Donovan, to liaise directly with Churchill.

'Big Bill' was a larger than life character; latterly a successful Wall Street lawyer, he had abandoned the brief for the sword at his president's request even though their politics were, in many areas, incompatible. A much decorated veteran of the Western Front, Donovan had maintained a finger in the counterintelligence pie and was a firm advocate of the Allied cause. He would go on to found the US equivalent to Special Operations Executive (SOE), the Office of Strategic Studies (OSS) which would, in due course during the emergence of the Cold War, grow into the omnipotent Central Intelligence Agency.

Having met and conferred with Churchill in London, Donovan proceeded on a grand tour of Eastern Europe, including the Bulgarian capital of Sofia, Belgrade, where he spoke with Prince Paul and then to North Africa for discussions with the representatives of Vichy.

His message to the War Cabinet was a simple one and typically forthright. If Britain wished to restore and maintain its foundering credibility with the Americans then some form of successful military venture had to be undertaken on the European mainland. Clearly France, the Low Countries and Scandinavia were beyond the Empire's much depleted resources; this left the Balkans, the deadly melting pot and graveyard of spent alliances.

It was this need, coupled with the spectre of an Italian offensive, that triggered Churchill's interest in Greece. If Britain, for a limited military commitment, could facilitate the defeat of Il Duce's strutting legions then this might achieve the necessary resonance across the Atlantic.

Mussolini's designs were scarcely secret. Since the annexation of Albania in 1939, the Italians shared a common border with the Greeks and there had been an escalating tide of provocation since. In addition to a series of stage-managed border 'incidents', the Greek cruiser *Helle* had been brazenly torpedoed in an utterly unprovoked attack.

Hitler had met his Italian counterpart in October 1940 at the Brenner Pass in the wake of the German defeat over the skies of southern England. Mussolini was in high good humour as their talk ranged over a sweeping agenda. He might have been less jovial had he realised that the Führer had already taken the decision to beef up the German presence in Romania, to the extent that the whole country became an occupied territory; a key shift of policy he chose not to share. Some two weeks prior to the summit Hitler had issued a directive setting out the defined objectives:

> To the world their [the military mission's] tasks will be to guide friendly Romania in organising and instructing her forces.
>
> The real tasks – which must not become apparent either to the Romanians or to our own troops – will be:
>
> To protect the oil district ...
>
> To prepare for deployment from Romanian bases ... in case a war with Soviet Russia is forced upon us.[15]

Il Duce roared with impotent rage when he finally learnt of the order and he blazed that he would behave in like manner – that Hitler would read of his projected invasion of Greece in the newspapers! He then gave orders for the invasion plans to be put in hand immediately.

Consequently General Metaxas, de facto dictator, was awoken in the early hours of 29 October 1940 to be presented with an Italian ultimatum. He was accused by the ambassador, Count Grazzi, of aiding and abetting the British, Italy's enemy – as quid pro quo Mussolini now demanded free access to and passage over Greece sovereign territory for his troops. Anything less than complete acqui-

escence would be considered an act of war. Inevitably, as anticipated, Metaxas refused to see his country so summarily shorn of
nationhood. Within a couple of hours Italian troops had crossed the
frontier.

That same morning Hitler arrived for a further summit in Florence
to be met by Mussolini who, grinning like a mountebank, promptly
postured 'Führer, we march.'

It was at this point that the Battle for Crete most probably became
inevitable.

CHAPTER 2

A New Thermopylae

As Mr. Churchill stated in his review of the campaign, the military authorities considered that there was a line which, given certain circumstances, could be successfully defended. The Greek campaign was not undertaken as a hopeless or suicidal operation. It turned out to be a rearguard action only...[1]

Given certain circumstances; the Official History does not define what these circumstances might have been and there has, with the inestimable benefit of hindsight, been a common perception amongst historians that the Greek adventure was a hare-brained notion from the start:

> The decision to go to Greece was a political one, and from the point of view of a professional it was a military nonsense ... the diversion of resources to Greece including 6th and 7th Australian divisions, the New Zealand division, and part of the 2nd Army took away from General Wavell in Africa practically the whole of the fighting formations which were ready and equipped for operations, and therefore by going to Greece we endangered our entire position in the Middle East.[2]

General Archibald Wavell, the Commander-in-Chief of British and Imperial forces in the Middle East, was far from sanguine about Allied prospects in Greece. He was rightly concerned that the Italian build up in Libya where Marshall Graziani's huge army dwarfed his own, represented the major threat. The General was, whilst a consummate professional, not imbued with the gift for dealing with awkward political masters. He seems to have, unfairly, been held in low esteem by Churchill whose swashbuckling approach, often totally unrelated to logistical constraints, was at odds with Wavell's caution and natural reserve.

Under his command he had some 70,000 troops, rather ill assorted and, with two of his potentially best contingents, the Australians, under General Thomas Blamey and the New Zealanders, under General Bernard Freyberg, reporting to their national governments rather than directly to the commander-in-chief.

The Italians had some 280,000 men, supported by 1,500 aircraft; Wavell's RAF support, headed by Sir Arthur Longmore, could barely muster 200 airworthy machines, many of these obsolete. Supply was by convoy taking the long route around the Cape of Good Hope and through the Red Sea, prone to attack by swift Italian destroyers and submarines. An overland air supply route over the trackless wastes of the Sahara was opened from Takoradi, each run an epic in itself.

In his directive of 16 August 1940 the Prime Minister stressed the vital importance of defending Egypt. Wavell certainly would not demur but he identified the overriding need, not for men but matériel, aircraft, trucks and tanks. Quite correctly he had judged that the war in the desert would be one that was decided by firepower and mobility, supported by superiority in the air. The General's conclusions were accepted and the supply of equipment stepped up accordingly. Churchill's bold idea of sending a convoy through the Axis-infested waters of the western Mediterranean, whilst extremely risky, paid off.

In Greece a great surge of patriotic fervour, sufficient to unite the many disparate factions, even under the leadership of a despised autocrat such as Metaxas, rallied and took on the Italian invaders. Despite a shortage of just about everything and a haphazard supply chain, the ill-armed Greek conscripts swiftly brought Mussolini's seemingly irresistible juggernaut to an abrupt halt. The Italian troops were not equipped for an autumn campaign, their morale proved illusory and a determined counter-attack, launched in mid-November, began, very swiftly, to assume the proportions of a rout.

Although Hitler was quick to criticise his hapless ally for the severity of the defeat, he was not perhaps as opposed to the idea of a Balkan involvement as he might have appeared. The idea of a co-ordinated attack on Greece and an offensive in North Africa, aimed at the capture of Suez and thereby, imperilling Britain's entire position, was not unattractive.

As early as the summer of 1940 German planners at both OKH[3] and OKW[4], the top tiers of the Nazi command structure, had considered the possibility of supporting an Italian invasion of the Greek

mainland with a simultaneous airborne assault on Crete. This would
only be launched when Graziani's legions had succeeded in capturing
Mersa Matruh in the second leg of the proposed desert offensive thus
providing the Axis with forward airstrips and bringing the British
fleet anchorage at Alexandria within bombing range.

The naval base at Souda Bay on the north coast of Crete, just east
of the island's administrative capital, Chania, would be an invaluable asset in the war at sea, one which was presently available to the
British. The scheme for proposed cooperation did not find favour
with Il Duce who saw his dreams of imperial conquest coming to
fruition purely as a result of Italian efforts, without the need for
German intermeddling. As General Franz Halder sourly remarked,
the Italians 'do not want us'.[5]

As the Germans were tentatively touting the idea of a combined
Balkans operation, in August the Greeks, already alarmed by Italian
sabre rattling and overt provocation, had approached the British
ambassador requesting assistance in the event, as now appeared
likely, of an invasion.

The subsequent report prepared by the Chiefs of Staff Committee
and delivered to the War Cabinet on 9 September was unequivocal:

> Even with the reinforcements at present contemplated, our land
> and air forces in the Middle East will be no more than sufficient
> to withstand a determined attack by Italian and German forces.
> Until the attack on Egypt has been finally defeated no forces will
> be available for assistance to Greece ... no forces can be made
> available for assistance to Greece until the present threat to
> Egypt has been liquidated.[6]

The overwhelming weight of military advice was therefore, from the
outset, against any intervention in Greece. Britain's stock of military
capital was simply too slender to face a fresh division of resources.
Despite having guaranteed Greek neutrality and despite the need to
impress the Americans, the generals both at home and in the Middle
East were opposed to sending troops.

The cold logic of this conclusion is inescapable, Britain had
suffered too many reverses and now faced a major threat to her
Empire. Greece must fight alone. Should the Germans intervene to
support their failing ally then the result would not be in doubt.
General Metaxas remained confident his raw levies could see off the
Italians but the advance of the Panzers would finish them.

Churchill, writing later, and with a fine eye for the useful benefits of hindsight, gives the wider political view:

> We often hear military experts inculcate the doctrine of giving priority to the decisive theatre. There is a lot in this. But in war, this principle, like all others, is governed by facts and circumstances; otherwise strategy would be too easy. It would become a drill book and not an art; it would depend upon rules, and not on an ever changing scene.[7]

1940 had been a year of defeats for Britain, her battered defences bolstered only by those brave fighters who had escaped from the occupied territories. Despite this, the overall strategic position seemed set to improve. The Greeks were busily trouncing the Italians; it seemed as though Albania could fall and British Fleet Air Arm Swordfish torpedo bombers scored a signal success against Il Duce's navy at harbour in Taranto. In the second week of December Wavell unleashed his desert offensive and inflicted a series of dramatic defeats on Graziani's army, to the extent that the whole Italian position in North Africa began rapidly to unravel.[8]

The earlier confidence of the Axis leaders, preparing to act as receivers of a bankrupt British Empire, evaporated. In December both Bulgaria and Yugoslavia declined to join the Tripartite Alliance; worse the Russian attitude began to harden. In the course of a conference in Berlin, Molotov displayed both caution and suspicion, even presuming to grill the Führer himself. A timely air raid by the RAF served to underline the fact of Britain's continuing defiance.

When Stalin later confirmed his foreign minister's demands and intransigence Hitler realised that the time for 'Barbarossa'[9] was at hand. From now on German strategic considerations were to be driven by the need to plan and prepare for the invasion of Russia; the Balkans would thus be a sideshow only, a necessary sideshow and one which had the advantage of providing a cover for the German build up.

The collapse of the Italian position in both Greece and North Africa removed any argument over the question of German involvement. Hitler had to secure this southern flank before committing his forces to an all out attack on the Soviets. Thus, the Wehrmacht was being drawn into a Balkan campaign not by policy but as a consequence of diplomatic failure and, on Il Duce's part, military bungling. After a further, unsuccessful effort to interest Franco in an attack on Gibraltar, Hitler was convinced of the need to intervene in Greece.

Churchill also had a fascination with the Balkans and the autumn successes against the Italians fuelled his interest in Greece where it might yet be possible to strike a blow against the Axis; such a blow as might serve to significantly bolster American enthusiasm for the Allied cause. The Prime Minister had convinced the Chiefs of Staff that a descent upon Rhodes, Mussolini's last Aegean bastion, might be accomplished by a mixed force of marines and commandos. In the event, this did not proceed.[10]

The idea of providing support for Greece, a small, poor nation that was fighting heroically and with success against the weaker arm of the Axis, was naturally appealing and carried the weight of moral imperative. Alan Clark quite rightly points out that, in the first four years of war, Greece was the only European power to defeat the fascists on the continent. The political will was therefore present but could the necessary military resources, planning and deployment be achieved?

When the Italians launched their invasion, Anthony Eden, British Foreign Secretary, was already in Cairo where Churchill badgered him with the feasibility of providing military aid. At this stage Eden appeared to share Wavell's view that such a division of resources would fatally weaken the planned offensive in the western desert. Only partially deflected, the Prime Minister then issued orders for further aircraft to be sent. This was done without the knowledge or acquiescence of Middle East Command and seriously weakened the RAF's position overall.

The Commander-in-Chief had already reiterated his earlier misgivings in a cable dated 2 November:

> As hostilities develop between Italy and Greece we must expect further, persistent calls for aid. It seems essential that we should be clear in our minds on this main issue now. We cannot from Middle East resources send sufficient air or land reinforcements to have any decisive influence on the course of the fighting in Greece. To send such forces from here or to divert reinforcements now on their way or approved would imperil our whole position in the Middle East and jeopardise plans for offensive operations. It would surely be bad strategy to allow ourselves to be diverted from this task and unwise to employ our forces in fragments in a theatre of war where they cannot be decisive....[and later in a

further communication of the following day]...in general all Commanders-in-Chief are strongly of the opinion that the defence of Egypt is of paramount importance to our whole position in the Middle East. They consider that from the strategical point of view the security of Egypt is the most urgent commitment and must take precedence of attempts to prevent Greece being overrun.'[11]

Buoyed by the sweeping successes of the December offensive, Eden, by January 1941, appeared far more willing to consider a deployment in the Balkans. Indeed Churchill now felt the Italians should not be pursued beyond the vital bastion of Tobruk and that, with the campaign, as he saw it, virtually over, a view that was supported by the South African General Smuts, attention could revert to the Balkans.

Wavell remained obdurate but he based his objections on the view that the German threat was more apparent than real. He might have been wiser to simply reiterate that Britain lacked the necessary resources to embark on a Greek expedition. This flaw earned an immediate rebuke from Churchill who sharply reminded the Commander-in-Chief that his obligations were to carry out orders rather than determine policy. Thus chastened, Wavell seems to have taken the view that he had no choice but to comply with his instructions regardless of his misgivings.

Consequently both he and Air Marshal Longmore were dispatched to Athens for a high level meeting with the Greek commanders. Wavell would not be disappointed to discover that the Commander-in-Chief, General Papagos, was far from enamoured of the concept of British intervention. He believed that a modest deployment of troops would achieve nothing and possibly only worsen the situation. Wavell was thus able to report a negative outcome and press on with his plans, already in hand, to pursue the beaten Italians as far as Benghazi.

Though Churchill concurred, he did not abandon the Greek venture and the sudden death of General Metaxas, who succumbed to a heart attack on 29 January, raised fresh possibilities. Anthony Eden, with the zeal of the convert and a desire to please, conceived the notion of a grand Balkan alliance, a united front that would be strong enough to thwart any Axis attempt to intervene further in Greece, who would join with Britain, Turkey and Yugoslavia.

If such a grouping could be brought into being then the number of divisions available would outnumber those which the Wehrmacht could hope to deploy in the Balkan sector. Noble in concept this grand strategy disregarded both political and military realities. Divisions which might exist on paper would seldom materialise in the flesh and the morass of Balkan politics generally proved bottomless.

On 8 February the Greek dictator's successor M. Koryzis requested further discussions on the size and nature of the British commitment. Wavell's cautious approach that the soundest policy was to continue naval and air support, while retaining Crete as a vital bastion but avoiding a large scale deployment, had been echoed by the Joint Planning Staff. The crushing weight of the British victory at Beda Fomm had reinforced Eden's enthusiasm for his Balkan Crusade.

If the Turks were to be the eastern bastion with the Yugoslavs to the north then it was essential that the Greeks, in the vulnerable centre, were speedily and substantially reinforced. A fresh mission to Athens would therefore be necessary, this time headed by Eden rather than Wavell or Sir John Dill, the Chief of the Imperial General Staff; a shift of emphasis from the purely military to the overtly political.

Even so Churchill, when writing to Wavell, intimated that if it was not possible to reach a working brief with the Greeks then all that could be done would be to salvage Crete and any other island bases which might be useful. When Eden met with Wavell, Dill, Longmore and Admiral Cunningham (Commander-in-Chief of the Mediterranean Fleet) in Cairo he once again found the General sceptical of the prospects for offensive military operations in the Balkan sector.

Eden was not easily dissuaded and wrung a concession from the C.-in-C. that a scheme for the defence of Salonika be devised and put to the Greeks. Cunningham later noted that both he and Dill saw little to offer in this but held their counsel.

Salonika, Greece's second city and her great northern port, one which had seen much Allied military activity in the Great War, was the pivot upon which the proposed Balkan alliance must turn; it was also the only harbour other than Piraeus that could provide a base for an expeditionary force. To hold the city however, the Greeks would have to base their defensive line on the northern chain of Macedonian passes. Wavell privately felt that this would be a good deal further than they would wish to extend and might, hopefully, allow the whole idea to founder.

He also expressed the clear need to have the Balkan allies on side before seeking a deployment around Salonika for, as matters stood, the Axis could invade long before any sufficient build up could be established and simply overrun the bridgehead. Wavell, who had already received a severe rebuff from Churchill, had also, even in the hour of his triumph in the desert, been subjected to a series of rather petty snubs and recurrent sniping. It was clear to him that he did not enjoy the Prime Minister's confidence and that his strategic assessments, invariably sound, were likely to be disregarded.

Eden, convinced the proposed intervention could be made real, did not feel constrained by such practical considerations. He conceded that the whole strategy constituted a 'gamble' but the risk was justified by the paramount need to be seen to be doing something to assist the Greeks. So fired was the Foreign Secretary by the urgency of his mission that, as his delegation flew on to Athens, there was some hasty and disreputable shuffling of the resources Britain was able to commit.[12]

At the Tatoi Conference and, after some considerable haggling and a further inflation of the figures, a form of accord was attained, though, as it transpired the Greeks were as nimble as the British in overestimating numbers. General Papagos was in favour of holding the northern line and thus securing Salonika but this would require the disengagement of troops currently deployed in Albania; a difficult and uncertain matter.

Churchill, by 7 April, appeared to be having second thoughts and, seeking to curb his Foreign Secretary's enthusiasm, cautioned that there was nothing to be gained from encouraging the Greeks to a doomed struggle if we had only troops in penny packets to offer.

A further complication now arose in that some 80 per cent of the forces to be deployed in Greece were to be drawn from the Australian and New Zealand contingents but these were not directly under the orders of the C.-in-C. Middle East nor, for that matter, the War Cabinet. Acceptance of the strategic reasons for the deployment should, according to protocol, be sought beforehand from the Dominion governments.

In the event both Blamey and Freyberg were simply given their orders without reference to Canberra or Wellington. This highhandedness was to spark understandable resentment in the wake of the Greek and later the Cretan debacle. The official reasoning, outlined in a cable sent to the acting Prime Minister in Australia, was the need to form Eden's hoped for Balkan Front.

The Greek Expedition suffered from the outset from a confusion of objectives and a raft of heroic assumptions which largely proved untenable. Both Eden and Papagos had wildly overestimated the numbers which would be available; the entire scheme was based on the concept of a Balkan Alliance which did not even begin, at this stage, to exist and was put forward in the teeth of opposition from all of the commanders involved – army, navy and air force. The decision to assist Greece was, from the outset, political rather than military. In practical terms the ability of either Britain or Greece being able to deliver the resources needed was, at best, highly questionable.

The question of where best to stand on the defensive was discussed at the Tatoi Conference. Here the bland eloquence of political assurance began to founder against the harsh reality of the topography. The first position was a line drawn along the Bulgarian Frontier which would safeguard Salonika or a rearward position buttressed by the slopes of Mount Olympus and the Vermion range – the Aliakmon Line stronger but, being some forty miles behind the first, would mean the abandonment of Salonika. The Greeks, for understandable patriotic reasons, wished to hold the frontier and deny the Germans the soil for which they had already fought so hard.

But to hold this it was necessary to have the support of Yugoslavia, the second bastion of Eden's proposed alliance. Militarily the British generals favoured the Aliakmon Line as the advanced position could easily be outflanked if the Germans attacked through Serb territory; a real possibility as the attitude of the Belgrade Government had yet to be ascertained. Despite the apparent confidence some of those present at the conference (as Colonel de Guingand confirms) were by no means sanguine at the prospects of holding the Aliakmon Line:

We had a momentous conference at Tatoi, the King of Greece sitting at the head of the table with his prime minister on one side, and Papagos, his commander-in-chief on the other. And I was actually there when Eden asked Dill to inform the King of Greece and his cabinet his views as to whether we'd be successful if we intervened in Greece, and Dill got up and made a most optimistic statement that he felt we could hold the line in Northern Greece called the Aliakmon Line and prevent the Germans from getting deep into Greece. I remember I was absolutely shattered because all our own studies in the ground planning staff had shown that it wasn't possible, you'd never get the forces in sufficient strength there in time before the Germans would be there, in sufficient strength to come right down into Greece.[13]

One of the main weaknesses of the strategy determined at Tatoi was the reliance on the Yugoslavs at a time when the mood in Belgrade was unknown, as indeed was the view in Ankara, the third capital in Eden's three great pillars. The Turks, whilst conciliatory, were not easily drawn. They had no reason to invite German aggression and any vague assurances were clearly dependent upon them receiving quantities of aircraft and matériel which Britain was not placed to supply.

The situation in Yugoslavia was even more uncertain; the country was a political creation, born of the dismemberment of Austria-Hungary after 1918. The uneasy mix of peoples was dominated by the Serbs who leaned toward Britain, their ally from the First War. There was, however, in Croatia a substantial minority who leaned toward Germany and, in February 1941, Hitler had made it clear to the Yugoslavs that he expected them to ally themselves unequivocally with the Axis.

Prince Paul, the Regent, treading a delicate path between the two protagonists, was inclined to accept the German accord with the assurance that Italy would not benefit at his country's expense. He was cautious as he feared too overt a move toward the Axis could produce a backlash that would unseat his government.

The Germans were not minded to temporise – it was a question of whether the Regent preferred an alliance or an occupation. At the same time he was fending off repeated calls from Britain with the result that his country stood unhappily poised in a continuing dichotomy.

The net result of this feverish diplomacy was entirely negative. Sir Anthony Eden's dream of a buttressed Balkan coalition was exposed as a chimera and the viability of the British military expedition to Greece fatally undermined before the first shots were fired. In London the War Cabinet was becoming alarmed, Eden's initiatives had been launched on the Prime Minister's sole authority but the bulldog refused to be cornered.

Admitting that the expedition might prove a military blunder, Churchill challenged his colleagues with the need, imperative as he now saw it, to support the Greeks regardless of loss. The Cabinet backed off but with the important proviso that the expedition must receive the full endorsement of the Dominion governments.

In the event, on 26 February the New Zealand administration concurred, buoyed by assurances from General Freyberg. It may certainly be said that the facts, particularly the prospects for Yugoslav and Turkish involvement, may have been cast in a singularly optimistic light. R.G. Menzies, the Australian Prime Minister who had been present at the cabinet meeting, also urged agreement though he seems to have felt a greater degree of unease than this urging would suggest. He commented that the decision was being undertaken on the basis of assessment supplied from Middle East Command who had, in fact, expressed grave reservations about the whole scheme.

Some authors have argued that the concurrence of the Dominion governments was gained 'by a combination of misunderstanding, misleading information, and straight untruth'.[14] This may be unduly censorious and it is likely that Churchill adopted the same steamroller tactics as he did with his own colleagues. The Axis dictators had the considerable advantage of being accorded demigod status and were not troubled by having to cajole members drawn from an elected assembly and with differing political hues. Goering, in March 1941, observed with practised sycophancy that 'The Führer is a unique leader, a gift of God. The rest of us can only fall in behind.' [15]

Churchill was a brilliant bully who was perfectly prepared to mislead or disregard in order to achieve his objectives. He was by no means infallible, great men seldom are, and, as Tony Simpson also mentions, the Dominion governments were perhaps still somewhat in awe of the mother parliament.

This is not to their discredit; the men who served from Australia and New Zealand were volunteers who believed in the truth of their cause and who were responding to a deep patriotism for Britain and the British way. This had perhaps been tarnished by Gallipoli and the subsequent bloody campaigns in France and in Flanders but it had not been extinguished. The Dominion governments responded because they saw the stark reality of Britain's peril and they therefore accepted that blood would be shed to secure freedom from Nazi tyranny in Europe. They may have been in part naive but they were not wrong.

By 13 December 1940 Hitler was outlining his plans for a Balkans campaign; this would begin in March 1941 and be expected to last no more than three weeks. Timing was everything for the divisions would soon be offered fresh employment elsewhere. The invasion of

mainland Greece and the occupation of Bulgaria, codenamed
'Marita' would be an exercise intended to secure the southern flank
while the main issue was settled on the Russian Steppe. Even the
deployment of General Rommel and his Afrika Korps expedition to
Libya was merely intended to bolster the Italians and keep the British
engaged rather than advance into Egypt and capture Suez. As Halder
noted in February: 'The war in Africa need not bother us very much
... but we must not risk the internal collapse of Italy, Italy must be
saved from that. It will be necessary to send some help.'[16]

A dramatic development occurred on 26 March 1941 when an
army coup unseated the Yugoslav Regent, took control of the person
of the young King Peter II and established a Serbian dominated
military regime. This course of events had been in part instigated by
Big Bill Donovan who had tapped into Serbian Nationalist, anti Axis
sentiment in Belgrade and the key garrisons.

Although the junta leaned now toward the Allies, the generals were
not so foolhardy as to hazard their tenuous grip on power by defying
Germany. Although there were discussions with General Papagos
early in April, these broke up in confusion but the die was already
cast. Outraged at what he perceived as standard Balkan duplicity
Hitler had, on 27 March, issued orders for the aptly named
Operation Punishment.

It was now the turn of Yugoslavia to experience the full horrors of
blitzkrieg with her air force shot to pieces on the ground and her
capital subjected to a murderous aerial bombardment that left the
city transformed into rubble and 17,000 of her citizens dead in the
ruins. The Yugoslav army had disintegrated even before the panzers
arrived and, on the morning of Sunday 6 April, five full armoured
divisions under General von List crossed the Greek frontier, together
with two motorised, three mountain, eight infantry and the *SS Adolf
Hitler* divisions. There was no Balkan Alliance but there was now a
Balkan War.

> No one expected the Greek campaign to be anything but a
> disaster. Long before any official announcement was made it was
> known we had troops in Greece, and I could find no one of
> whatever kind who believed that the expedition would be suc-
> cessful; on the other hand nearly everyone felt it was our duty to
> intervene. It is generally recognised that as yet we can't fight the
> Germans on the continent of Europe but at the same time 'we
> couldn't let the Greeks down'.[17]

As Salonika was too exposed for disembarkation, the majority of British and Dominion troops came ashore at Piraeus or further north at Volos which was closer to the forward post at Larissa. In total the forces dispatched totalled some 58,000 men, of whom roughly 35,000 were front liners, with the rest support and administrative personnel. The Dominion divisions were between 10,000-15,000 strong at the outset and took with them their own divisional artillery, mainly the highly effective 25-pounder field gun and the considerably less useful 2-pounder anti-tank gun, obsolete and generally ineffective against panzers.

The Kiwis had their mechanised battalion equipped with light tanks and Bren carriers – the 'divisional cavalry regiment'. In addition to an anti-tank regiment the divisions were equipped with anti-tank rifles and Brens; the men carried .303 Lee-Enfield bolt action rifles as their personal weapons, together with a few Thompson sub-machine guns, the ubiquitous 'tommy gun', a .45 calibre weapon which was capable of firing in bursts. A large number of trucks was available and the single British armoured division of 3,000 - 4,000 soldiers had around 100 tanks, together with field artillery, anti-tank and engineer formations. It also possessed some anti-tank rifles and light mortars.

Nominally the Greek army could dispose some fourteen divisions in Albania and three and a half on the Bulgarian border. Inevitably some of these formations existed more in name than reality, were poorly equipped and their morale had been sapped by losses and the intensity of the campaign fought in the freezing passes. They possessed little modern artillery, less tanks and only the most basic logistics; most supplies came on the backs of mules or donkeys. They were totally exposed to attack from the air. From the outset the Luftwaffe enjoyed an almost unchallenged superiority in the air.

The lack of air cover was a constant nightmare for the hard-pressed and weary troops on the ground – much criticism was levelled at the RAF ('Rare as Fairies' and other epithets). This was not due to faintheartedness – the planes were simply not available; the crucial element for the success of any modern campaign, adequate air support, was lacking from the very start.

> We marched and groaned beneath our load,
> Whilst Jerry bombed us off the road,
> He chased us here, he chased us there,
> The bastards chased us everywhere.

And whilst he dropped his load of death,
We cursed the bloody RAF,
And when we heard the wireless news,
When portly Winston aired his views -
The RAF was now in Greece
Fighting hard to win the peace;
We scratched our heads and said "Pig's arse",
For this to us was just a farce,
For if in Greece the air force be -
Then where the bloody hell are we?[18]

The initial Allied plan was that the three Greek divisions, under-equipped, under-strength and under-supplied, would be used as a blocking force to blunt the German onslaught. The remainder of the available Greek forces were enmeshed with the three Italian armies operating in Albania. This attempt at a holding action was never really a viable proposition and the blow launched on 6 April across the Bulgarian border and to the east of Salonika, was delivered in overwhelming force, with full and close air support.

Paratroops were dropped behind the Greek lines guarding the Rupel Pass but this early deployment of airborne troops was not a success; most of the detachment of 150 were killed or captured as the Greeks fought back with considerable gallantry. Nonetheless Salonika fell within days.

Greece was, to all intents and purposes, a country with a near medieval infrastructure. A single railway line wound from Athens to Salonika, a narrow and highly vulnerable ribbon that connected the two principal cities. Roads were little more than tracks, unsuitable for motor vehicles and impassable in bad weather. The Allied commander, General 'Jumbo' Wilson, was further hamstrung by the fact that, in order to satisfy the Greeks, still officially neutral, he was obliged to pretend he did not really exist, masquerading as a jour-nalist!

As war between Germany and Greece had not, prior to 6 April, been declared, the German legation in Athens was not troubled and the military observers were able to observe without interference. It was into this almost Ruritanian atmosphere that British and Dominion troops were disembarking:

The Greeks hadn't declared war on Germany – it was an amazing thing. The fellows from the German embassy were quite openly walking about with us. There was a blackout in the town at night for aircraft and I remember going to the local night-spot a place called Maxime's – a sort of night club. And the fellows from the German Embassy were all there in civvies, drinking and laughing at us. There was a bit of trouble because one of our blokes got into the German Embassy and stole a pair of very expensive pyjamas. Nothing ever came of it to my knowledge because the Greeks declared war on Germany shortly thereafter.[19]

The situation had the makings of comic opera but the consequences of this extraordinary lax security were serious enough – by 9 March OKW in Berlin had a full and accurate assessment of Allied strength and intentions. If Wilson's problems were not sufficient, he struggled to exercise any proper form of command structure beset by logistical difficulties imposed by an unhealthy mix of Balkan politics, difficult terrain, poor communications and muddled objectives. These problems were further exacerbated by a lack of standardisation and cooperation between the forces at his disposal, each clinging rigidly to its own pre-determined structure.

Brigadier 'Bruno' Brunskill, to whom was passed the poisoned chalice of coordinating the logistical effort, found, initially, he was not able to move freely north of Larissa for fear of antagonising the Germans! His only reconnaissance was by air and he was obliged to rely on a single borrowed map, his efforts to blend Greek and Allied supply networks were doomed to failure.

Ultimately he was forced to fall back on the tried expedient of heroic improvisation; the Wehrmacht was not labouring under such intense difficulties and in the finest blitzkrieg tradition had been able to concentrate its attacking forces, the *Schwerpunkt*, in overwhelming strength in the right place and fully supported from the air. No improvisation, however brilliant or inspired, can triumph against such deadly precision.

To ensure that this superiority in the air was maintained, Von List was the beneficiary of a new aerial command, Luftflotte 4. A mix of fighters, bombers, dive bombers and observers with a full complement of aircrew and base personnel, created specifically to support the Balkans offensive, some 1,200 aircraft flying from bases in Romania and Bulgaria. Against this formidable armada, backed by a further 300 Italian planes, the RAF could, as mentioned, muster a paper strength of perhaps a couple of hundred, less than half of which were airworthy. It was an inauspicious beginning.

The primitive state of the Greek national infrastructure further hamstrung the Allies, whilst the Germans had been able to create functioning airstrips in the occupied countries. To signal the type of interference the British might expect, the Luftwaffe carried out a serious raid on the docks at Piraeus on the night of 7 April. One of their targets was the supply ship *Clan Fraser* – loaded with munitions. Brunskill, arriving by car to do what he could, found total chaos:

> To my dismay I saw the port was in flames. The fire on the *Clan Fraser* had taken such a hold there was no possibility of putting it out. There was not a Greek in sight nor any member of the crew. Red hot fragments from the ship had started fires wherever they dropped on buildings and more important on every ship, lighter and boat. No one seemed to be doing anything to save the ships. I found a small party of New Zealanders and we put out a few small fires with buckets of water.[20]

There was little that could be done. Presently a further ammunition supply vessel also blew up to add to the confusion and to reinforce the message that the Luftwaffe ruled the skies over Greece.

Despite all these difficulties the Allied soldiers found themselves being greeted rapturously by the citizens, welcomed as friends and liberators, an affection that never waned even when the same troops, battered, bloodied and in the exhaustion of retreat, came stumbling back through the same streets.

By the middle of March the battalions were digging in along the Aliakmon line but the plan was already crumbling. General Papagos was unwilling and largely unable to extricate his divisions from Albania, the newly raised formations were hopelessly inadequate and under-equipped. Wavell's expressed concerns over the vulnerability of the Allied defences proved entirely well founded. The line was also very thinly held and a vital corridor, through which the Germans could penetrate and thus turn the whole position, was virtually unmanned. Hitler was not blind to the strategic opportunities his sudden and violent occupation of Yugosalvia now presented.

There was no alternative but to withdraw and Wilson extricated his forces from the trap which the Aliakmon line had become to establish a new position which, in the east, would stretch from the anchor of Mount Olympus to the Serbian border. In the course of the withdrawal the Greek Macedonian divisions began to disintegrate; a whiff of treachery was also in the air.[21]

This proved to be the beginning of a series of extended rearguard actions into which the campaign deteriorated, faced with the continuous advance of an enemy with an overwhelming superiority of men, guns and armoured vehicles, his advance closely supported at every stage by the siren wail of the Stukas and the murderous strafing runs of Me109s.

Sir Lawrence Pumphrey had been commissioned into the Northumberland Hussars; Yeomanry, who were being re-equipped with anti-tank guns and obliged to part with their beloved mounts (officers had been allowed to retain their horses, however, until the close of the current hunting season). He arrived in Egypt on New Year's Day 1941. The regiment was encamped near Tel-el-Kebir, site of Sir Garnet Wolseley's victory over dissident local forces in 1882. Here they practised the techniques of desert warfare before moving closer to Alexandria, on the fringes of the Nile Delta.

Rumours abounded, and soon the 'Noodles' were aboard HMS *Gloucester* bound for Piraeus. Young men, schooled in the classical tradition and reared on Homer, strained for a first glimpse of the magical Greek mainland. Initially they were based at Glyfada, amidst delightful hill country cooled by thyme scented spring air.

Squadron Leader David Barnett, who would be killed on Crete on the morning of 20 May, gave a series of lectures before the unit moved into position on the line of the River Axios in late March. The Noodles were temporarily attached as gunnery support to the Greek 6th Division under General Karasos. Sir Lawrence found the General spoke no English, and they communicated with a little oral French and by written exchanges in classical Greek.

On 6 April, Easter Sunday, Sir Lawrence returned from communion to find the entire Greek force had decamped without notice. His detached four gun battery was obliged to limber hurriedly and follow the rest of the regiment in their retreat through Thessaly. German tanks were catching up on the far side of the intervening river and the guns made ready to stand and fight. Before the enemy was sighted, fresh orders to withdraw were received and the rearguard once again struggled after, losing one of the guns during the course of a difficult river crossing.

Sir Lawrence's battery was supported by a platoon from the Rangers and, even though they were bringing up the rear, never actually sighted the Germans. They did not rejoin the main body until they had reached Thermopylae; ground sanctified by the sacrifice of much earlier wars.

The line was always in imminent risk of being outflanked and the panzers probed around the extremities, avoiding strongly held defensive positions which could be encircled and mopped up later; the pressure from the air was nerve-racking, relentless and virtually unopposed:

> The Germans dive bombed the village and put the wind up us. Do not like the dive bombers or the machine gunning from the air. It seems like years since I took my clothes off, had a wash and some sleep ... We saw some great aerial displays by the Hun. He doesn't move unless his airforce is going flat out and he has the planes and he keeps our heads down. Some more of his infantry crossed the flat yesterday and we gave them a hot time. I only dropped one but it was a long shot. Am just about asleep on my feet and everything on me is wet through. Bottom of the trench a sea of mud.[22]

M. Koryzis, the Greek Prime Minister, on hearing from Papagos (as did Wilson) that the Greek armies had reached the limit of their endurance, chose the moment to end his political career by blowing his brains out. His commander-in-chief, now becoming anxious to spare his country further suffering in the face of inevitable capitulation, suggested to Wilson that it was time for the Allies to withdraw, *sauve qui peut*.

This was the news that Wavell must have dreaded; all his worst fears were confirmed and the War Cabinet accepted the inevitable, endorsing the order for an evacuation; it was now just a question of how many could be saved from the gathering debacle.

On the ground Wilson faced the unenviable task of attempting a fighting withdrawal from the ruptured position around Mount Olympus to a shorter line of no more than fifty miles and running from the heroic outpost at Thermopylae to the Gulf of Corinth.

A nightmare retreat for the troops, dodging from one bomb-racked village to another, constantly exposed to the chattering machine guns, the trucks bumping and toiling over unmade cart tracks, exhausted, disorientated. All movement was confined to the hours of darkness, the drivers straining to maintain contact with the tail-lights of the truck in front, the only illumination permitted, the stark mountain landscape lit by the bursting of enemy shells and the crump of abandoned supply dumps going up.

Even when they reached their new positions the Allies were as exposed as ever. Wavell now flew to Athens to take stock of the unfolding disaster and to confer with both Wilson and the Greeks. The King was already preparing to evacuate his court and entourage to Crete and, after receiving their Commander-in-Chief's pessimistic report of 21 April, the War Cabinet confirmed the order to withdraw. It was none too soon for the Thermopylae line was now looking untenable; many units had lost much or all of their artillery and anti-tank weapons, and casualties, principally caused by air attacks, were mounting.

Evacuation, in the teeth of German hegemony in the skies was problematic, Piraeus was impractical because of this which meant the withdrawal would have to be accomplished through the necklace of small harbours further south in the Peloponnese. All heavy equipment would have to be rendered useless and abandoned, movement for the fighting formations was only permitted at night with the troops filtering down to their evacuation points, non combatant units had to take their chances during the lengthening spring days; easy meat for the prowling Stukas.

Sir Lawrence Pumphrey, and the Noodles, had bypassed Athens as they withdrew under the cover of darkness and at Marathon experienced the demoralising chore of disabling their surviving vehicles and guns. From Ruffina they were taken off by HMS *Fearless*. Though they had been through the whole campaign, they had not once sighted the enemy, other than the prowling Stukas and Me109s.

On 22 April the Greeks formally surrendered and by the 30th the evacuation by sea from the beaches was largely complete, despite a successful attempt by German paratroops to seize a vital crossing at Corinth by a coup de main. The Navy, not for the first, or last, time had delivered the rump of the army, some 80 per cent, from certain death or capture. Behind them the defeated army left all of their vehicles, heavy guns, armour and anti-tank weapons with great quantities of small arms, spares and supplies. Brigadier Brunskill recalled the moment of his deliverance:

Towards sunset [on 24 April] we were told to be ready to march. Relief showed on all our faces and no one asked the distance. The six miles were covered in excellent time and we arrived at the water's edge in good order without incident. As the harbour was too shallow for huge ships to put alongside invasion barges were brought in by the Navy and everyone looked on in silent admiration of real efficiency. The last on board arrived about midnight making seventy sleepless hours in all, and lay down everywhere, anyhow to sleep the sleep of exhaustion.[23]

Whilst the evacuation was a brilliant operation, superbly handled by the Navy, it remained a perilous operation; the departing ships, once the sun had risen, exposed to the fury of the Luftwaffe. The captain of HMS *Calcutta* graphically describes the intensity of the aerial harassment:

> At seven o' clock in the morning, April 27th, bombers came over and did not leave us until 10 a.m. We were shooting so accurately that again and again we put them off. About 7.15 one transport was hit and began sinking. I ordered the *Diamond* alongside to take off troops, and about 9 a.m. three more destroyers, the *Wryneck*, *Vampire* and *Voyager*, joined us in the battle with the dive bombers. In that three hours the *Calcutta* fired about 1200 rounds of four-inch shells and many thousands of rounds of pom pom and machine gun ammunition. The *Coventry* came out to relieve me, enabling me to disembark them and return to the convoy in the afternoon. One more transport was sunk but we got all the survivors safely ashore.[24]

The news from the Western Desert where Rommel and his Afrika Korps were making their presence felt was also disheartening; a lightening strike, launched on 24 March, had retaken Benghazi and was rushing eastwards toward Egypt. Wavell's stock plummeted and it was inevitable he would be selected as the chosen scapegoat for the Greek disaster.

At 5.54 p.m. on 25 April some 5,000 men from the 19 Brigade were landed at Souda Bay; the first contingent of Wilson's battered evacuees, '... They had very little in the way of arms or personal equipment; they were dirty, ill organised, with no proper chain of command existing, "bomb-shy" and conscious of their recent defeat.'[25] Crete; the backwater, ill manned, barely considered, was about to become the new front line.

CHAPTER 3

Operation Mercury

We are few yet our blood is wild,
Dread neither foe nor death
One thing we know – for Germany in need – we care
We fight, we win, we die,
To arms! To arms!
There's no way back, no way back.[1]

General Kurt Student was a man with a mission. His objective was to ensure the capture of Crete, which now began to enter the strategic arena, was accomplished by the parachute troops under his command, the elite *Fallschirmjäger* in what would be the ultimate test of airborne operations; 'vertical envelopment' from the skies.

Parachutists were a relatively new phenomenon in warfare, those who had slogged through the blood soaked horrors of the trenches had witnessed the birth of air power. It was a logical development of larger transport planes in the twenties that raised the possibility of these being used to outflank enemy positions from the air and drop infantry behind the lines to seize key objectives.

This delivery of men and matériel could be managed either by parachute or by glider or, latterly, by a judicious combination of both. The advantage of dropping from the skies was that the parachutists could land virtually anywhere and, as it would be hoped, muster swiftly on the ground to maximise the advantage of surprise.

The abiding weakness was that it was not possible to drop heavy weapons; these could only be accommodated by the gliders. The obvious role for airborne forces was to seize key objectives behind the enemy lines, interdict or sabotage his communications and be able to 'hang on' until relieved by their advancing infantry.

An additional and more ambitious use was for the parachutists and glider borne troops to be dropped some way behind the enemy's line with a view to attaining a more ambitious strategic objective. For this, maximum support from friendly aircraft for tactical firepower and re-supply was essential. The descent upon Crete would be just such an operation, the most ambitious yet undertaken by an airborne spearhead.

Soviet Russia was the first of the great powers to experiment with paratroops; as early as 1930 airborne contingents were featuring in exercises but the doctrine went into decline when its champion, Marshal Tukhachevsky, fell victim to one of Stalin's relentless internecine purges.

Germany had, however, watched developments by the Red Army with interest; both cooperated in the 1920s and 1930s in accordance with the terms of the 1922 Treaty of Rapallo. The restriction on German military expansion, imposed under the terms of the Treaty of Versailles and which had fettered the Weimar Republic, was repudiated by Hitler in 1935. This allowed the introduction of the Ju52/3^2 as a suitable workhorse for the Luftwaffe.

Major F.W. Immanns was the first commander of the nascent airborne arm in 1936, initially under the control of the army rather than the air force. The type of parachute adopted was the back pack version opened by a static line. In the course of the following year a formal training school was set up which also trained personnel from the Luftwaffe; not to be outdone the SS and even the SA sent odd detachments for training.

Immanns' successor, Colonel Bassenge, found that no coherent doctrine was emerging with men being trained in penny packets from rival services. Despite a limited showing in the 1937 manoeuvres, a viable parachute arm seemed unlikely until Student was appointed as operational commander.

A career soldier and consummate professional, Student had begun his career as a lieutenant in a Jäger battalion before the Great War. It was in the air rather than on the ground that he earned his reputation as a determined and fearless warrior, flying first against the Russians and latterly over the Western Front in the deadly circling dogfights fought out over the web of trenches. Shot down and badly wounded in 1917 he survived the war but saw the lustre of his profession dimmed by the savage repression of Versailles.

Student, the heroic flying ace, was not long unemployed, joining the fledgling and semi clandestine cadre of officers responsible for keeping alive the spirit of an air force despite the restrictions imposed by the Allies. With the emergence of Hitler and the triumph of

National Socialism in 1933, the chains were cast off and the Luftwaffe was set up under Goering, another celebrated fighter pilot. Whilst he showed little inclination to politics and did not hunger for the laurels and spoils of patronage, Student was ambitious, extremely able and, in the pursuit of his goals, ruthless.

His aloof manner and intellectual arrogance won few friends amongst his conventionally minded colleagues though his care for the men who served under him, his fearlessness and dash, earned him the unstinting admiration of his subordinates. This was an officer who led from the front and showed an almost total disregard for danger:

> [He] had absolutely no fear of danger ... driving with a very noticeable vehicle in areas held by partisans, in cities occupied by the enemy or in terrain dominated by enemy bombing – never with any security precautions. He paid no attention to random shots that flew around and seemed to be surprised when those who were with him threw themselves under cover. He wanted to give a visible example. This naturally made an impression on [his] parachutists.[3]

Student was able to cultivate a special relationship with Hermann Goering - the idea that airborne operations should be entirely the preserve of the Luftwaffe appealed to his mania for self-aggrandise-ment. Having established a corridor to the seat of authority and resources Student, formally appointed on 4 June 1938, set to work with a will.

His first task was to weld the various fledgling detachments into a whole, the *7th Flieger* division of the Luftwaffe. The development of the type DFS 230 glider[4] provided just the type of aircraft necessary to facilitate the creation of a glider borne arm. Although the Sudeten crisis of 1938 was averted by frantic diplomacy, Student took the chance to mount a full-scale exercise deploying 250 Ju52s over open ground. A carefully staged performance which, whilst it impressed Goering, failed to make an equal impression on the General's more sceptical Wehrmacht colleagues, already tainted by jealousy at his easy access to the fat *Reichsmarschall*.

The army withdrew its personnel from the airborne division which was left bereft of men. Undaunted, Student continued to preach his tactical doctrine and when control of the rump of the airborne arm came under full Luftwaffe control at the end of the year, it seemed his moment had come. With a second *Fallschirmjäger* regiment being raised in 1939, Student perfected his blueprint for airborne operations. His elite would operate on the ground like a conventional brigade with integral signals, mortar and light artillery.[5]

From the beginning admission into the ranks of the parachutists conferred special status. These young recruits were subject to a gruelling training regime and frequently came through the Hitler Youth, the very ideal of the Aryan warrior, fit descendants of Teutonic Knights and torch bearers of the Nazi ideology.

These young lions were marked by their distinctive uniforms, originally a blue-grey coverall over which they wore a rush green smock with zip breast pockets. This reached to just below the knee but could be fastened up around the upper thighs so as to prevent the parachute harness from fouling. With their trousers tucked into black calf length jump boots and their equally unique round padded helmets, every facet of their appearance marked them as an elite.

Specialist troops required specialist weapons and whilst many *Fallschirmjäger* went into battle carrying the standard infantry rifle, the KAR-98K, 7.9 mm and with a five round box magazine, others carried the MP-40 machine pistol, a 9 mm weapon with a 32 round magazine. This, developed from the earlier MP-38, was primarily designed as a paratroopers' weapon; light and with a high rate of fire it was ideal for airborne operations.

Additional and heavier fire support was provided by the 7.92 mm MG 34. Defined as a light machine gun, but extremely well designed, robust and versatile, it could be used by one man as a section support or by a crew of three as a medium machine gun. It could fire 800 - 900 rounds per minute.

All small arms and machine guns were packed in containers for the jump and it was vital the troops accessed these as soon after landing as possible. When they left the plane the individual paratroops carried only a fighting knife and a 9 mm Walther P-38 semi-automatic pistol.

Their British adversaries, encountering these elite warriors for the first time on Crete, were impressed:

Superbly equipped, on the whole elite troops, they were young, they were fit, they had brains, military brains, which is not as dismissive as it may sound, and their morale was terrific, they were very good soldiers ... They had some sort of outer garment like a kind of mackintosh which they got rid of as quickly as

possible and they were in an all purpose uniform with pockets
and fasteners – a very advanced looking battle-dress. They had
pockets for carrying their magazines, for instance. Someone had
obviously thought out the function of a paratrooper, how they
should be dressed in every conceivable detail had clearly been
gone into.[6]

Student would have been gratified, his men were imbued with the
ethos of Teutonic mastery, expected to adhere to a fierce 'moral' code
whose strident tone smacks of the days of chivalry when their mailed
ancestors had stamped their presence on the East Prussian landscape:

> You are the chosen ones of the German army. You will seek
> combat and train yourself to endure any manner of test. To you
> the battle shall be fulfilment. Cultivate true comradeship, for by
> the aid of your comrades you will conquer or die ... Tune
> yourself to the topmost pitch. Be as nimble as a greyhound, as
> tough as leather, as hard as Krupp steel, and so you shall be the
> German warrior incarnate.[7]

The lightening victory over Poland in the autumn of 1939 moved
with such dazzling rapidity that no opportunity for the *Flieger*
division to demonstrate its mettle arose. It was only with the subsequent campaigns in the west, beginning with the invasions of
Norway and Denmark, that Student finally found his chance.

Operation Weserubung was to involve the 1st Battalion, 1
Parachute Regiment under Captain Erich Walter. The
Fallschirmjäger were tasked to support the seaborne invasions of
both countries by the seizure of certain key objectives; in Norway,
Oslo and Stavanger airfields, two further airstrips at Aalborg in
Denmark and the capture of a vital bridge at Copenhagen.

Despite adverse weather over Oslo the paratroops, though
dispersed, managed to win their objective; the other three were
attained without serious opposition. As the campaign in Norway
progressed a successful and daring landing on packed ice contributed
greatly to operations around Narvik. This was vindication indeed.

It was, however, during the larger campaigns in the west that
Student's paratroopers were to achieve their most stunning successes,
victories that would put Student into personal contact with Hitler
himself, a vast increase in prestige and the continuing spite of his less
favoured contemporaries.

The operational tasks assigned to parachutists comprised the seizure by a *coup de main* of the apparently impregnable Belgian fortress of Eben-Emael, together with three vital bridges over the Albert Canal. Even more ambitiously, the 22nd Division was to air-land around the Hague in a dramatic bid to capture the Dutch Royal family and seize a number of airfields. The *7th Flieger* division was to take and hold the crossing points necessary for the relief by the advancing Wehrmacht of 22nd Division.

Captain Walter Koch was given the command of *Sturmabteilung Koch*, responsible for the extremely difficult job of assaulting the fortress and bridges. This was to be a glider borne operation and eleven aircraft under Lieutenant Witzig were to land directly on the roof of the fort.

The mission was a dazzling success, two out of the three bridges were taken intact and the fort's defences sabotaged by Witzig's group, even though his glider failed to make the drop, landing inside Germany after the tow parted too soon. The demoralised Belgians surrendered Eben-Emael when the ground forces arrived; the strongest garrison in the west had been reduced by a mere handful of paratroops.

The larger scale landings in Holland, however, were less convincing and revealed the weaknesses of Student's theories. Paramount amongst these was the fact that it was nigh on impossible for Student himself acting as divisional commander of the *7th Flieger* division also to successfully coordinate the actions of the 22nd (under Lieutenant General Hans Graf von Sponeck) which was landed some distance away.

Determined Dutch resistance foiled the paratroops' attempts to seize and hold the airfields which would facilitate the air landing of the remainder of the division. Sponeck's forces were thus scattered and, at the same time, contained. Kesselring, commanding *Luftflotte 2*, correctly assessed the situation and ordered Sponeck to simply consolidate his forces and then break out toward Rotterdam and Student. Though abortive, the landing did disrupt and tie down Dutch forces in considerable numbers.

The *7th Flieger* division was to take three bridges at Moerdijk, Dordrecht and Rotterdam. In the first and third instance the attacks were completely successful but, in the centre, the paratroops succeeded in gaining only a foothold and this was lost to spirited counter-attacks, the German commander being amongst the casualties. Despite substantial local counter-attacks Student's men clung to their gains until relieved on 12 May by 7th Panzer.

One of the German casualties was Student himself whose bravery in the field nearly proved his undoing. He suffered a very serious head wound which kept him out of the action for months. Although he seemed to recover there were those who felt his capacity to command had been impaired.[8]

While Student was incapacitated his temporary replacement, General Putzier, was involved in the planning of Operation Seelöwe (Sealion) the projected invasion of Britain but the paratroop regiments were not destined for further action until Marita and an attempt to prevent the retreating Allies from gaining the ports on the Peloponnese.

Colonel Albert Sturm's 2 Parachute Regiment was tasked, on 26 April, to seize the bridge over the Corinth Canal. In the event this came too late to hamstring Wilson's army as hoped but the bridge was wired for demolition. A dawn descent by three DFS 230 gliders landed assault engineers who succeeded in seizing the bridge and set about defusing the explosives. The 1st and 2nd Battalions dropped north and south to hold off the defenders.

To the north the relatively few British were quickly overrun, though resistance to the south proved more formidable. In the heat of the fight a Bofors, firing on the engineers swarming over the bridge, set off an explosion which caused the main span to collapse dramatically into the Canal below. In one sense the operation was successful but German casualties at around 240, sixty-three of whom had been killed, had been high. Nonetheless Operation Hannibal as this was named, captured twenty-one officers and 900 other ranks of the British and Dominion forces and 1,450 Greeks.

Student, substantively recovered from his wounds, was furious – the operation had revealed the existence of paratroopers in Greece, something he had been at great pains to conceal. Surprise, the vital element, was therefore sacrificed and the General had now set his sights on Crete.

Here was the perfect opportunity for a definitive campaign of vertical envelopment, one which would guarantee Student's place in the Pantheon of strategic genius, silence his critics with a single blow and cement his relationships with both Goering and Hitler. It was a dazzling prospect.

> He jumps through the air with the greatest of ease,
> His feet are together and so are his knees.
> If his 'chute doesn't open, he'll fall like a stone,
> And we'll cart him away on a spoon.[9]

The parachutist is taken on a short flight directly from his base camp, and without any middle act is plunged straight into close combat with his adversary. Without reconnaissance, without close contact with other forces or formations, he jumps into absolutely unknown territory. The only support he can expect is from his own bombers which drop according to a rigid plan which may even endanger his own life ... he does not fight on a single front but on all sides. Fundamentally he starts fighting in a situation which most infantrymen would regard as hopeless, for he ventures voluntarily, without tanks or artillery, into a total encirclement.[10]

For Student the casual ease with which von List's panzers had brushed aside the Greeks and their British allies was a source of considerable frustration. Early in the planning stage for Marita, Crete had been identified as a potential target for airborne assault. As such, it was the perfect opportunity for the *Fallschirmjäger* to consolidate their reputation and silence any critics for good. For Student it was the gift he had been seeking, the chance to overcome a conventional defence in a campaign of vertical envelopment.

Unlike some of his more flamboyant colleagues Student was not hungry for the adulation of the masses through the popular press; he lacked the genius for self publicity that characterised generals such as Guderian or Rommel. His was a professional conceit, his own particular genius rebuked by the petty mutterings and scepticism of lesser intellects. He now sought to create his masterpiece and Crete was to be his canvas.

The call, however, did not come and Student, increasingly frustrated, remained on the sidelines while the battle for Greece unfolded. It is uncertain if he was, at this point, privy to the plans for Barbarossa. If so, then this would only serve to exacerbate his concern, for the airborne forces had not been allotted any specific role. If what promised to be the climactic campaign of the war offered nothing, then time was indeed short.

On 20 April Student took the bull by the horns and, trading heavily on his stock with Goering, flew to Semmering in Austria, the operational HQ for Marita. His request for an audience was granted and his proposal listened to. General Jeschonnek, Luftwaffe Chief of Staff, was present and it was he who had raised the possibility of an airborne assault on Crete and possibly Malta in February. Goering had also been cautioned by General Lohr of *Luftflotte 4* that to leave the British in possession of Crete would be to expose Axis operations in the Eastern Mediterranean to continuing interference at a time when this flank need to be nailed secure.

The *Reichsmarschall* listened attentively and managed to secure an interview with Hitler at short notice, normally an impossibility. Student and Jeschonnek therefore attended the Führer aboard his headquarters train, *Amerika*. Before they saw the supreme commander they had to convince Jodl and Keitel of OKW; conventional generals who dismissed the scheme as fanciful. They preferred Malta as a target for Student's paratroopers.

Hitler, however, was prepared to listen. Bombers from Crete could strike at the Romanian oilfields, more vital than ever once Russian supplies were terminated. The deciding factor was undoubtedly the notion that the operation against Crete could form part of an overall deception, masking the real reason for German build up in the Balkans. The attack could not only secure the southern flank but also assist in maintaining the element of surprise upon which the success of Barbarossa was dependent. Student's wider vision for a continuing series of airborne attacks on Cyprus and ultimately Suez, was of little interest to Hitler but the potential to facilitate Barbarossa won the day for the Luftwaffe.

A few days later, on 25 April, Hitler confirmed his intentions to proceed against Crete; Directive Number 28 – Operation Merkur (Mercury) was born. This was to be primarily a Luftwaffe operation with additional army units thrown in; some armour would be transported by sea, the convoys guarded by Italian warships. It must not, however, in any way, interfere with the concentration of resources being undertaken for the invasion of Russia. For Student, the campaign was to be a race against time as men and matériel were progressively sucked into the web of Barbarossa. He would be in no doubt that the Wehrmacht would be resentful and uncooperative.

A lesser commander might well have been daunted but Student, the triumph of Hitler's directive in his pocket, was determined to prove equal to the task. The logistical difficulties were formidable in the extreme. Men and their equipment had to be moved across the difficult terrain of the Balkans without interfering with the contrary flow intended for Barbarossa. The air and flak units he would need had to be ready for re-deployment in the east by the end of the following month. Halder, in conference at OKW on 12 May complained openly of the inconvenience caused by Mercury.

For Student, the meticulous planner, this meant a foray into the uncertain world of make do and mend. From the outset the campaign was a race against time. It was not the strength of the British opposition which concerned him but the demands of the looming campaign in Russia. He had staked everything on the success of Mercury – his own reputation, his special relationship with Goering, even the future of German airborne forces. In the modern vernacular – failure was not an option.

Suddenly the men of the *Flieger* division, left kicking their heels in barracks since the heady days of 1940, were on the move. They did not yet know their destination but the heady lure of action was at last in the air, dispelling the tedium and impotence of garrison life. Student's competent quartermaster, General Seibt, had worked out an elaborate schedule to ferry the men to assembly points in Romania from where they'd pile into trucks that would carry them to their forward bases in Greece. The gliders required for the operation were crated and sent by train to Skopje where they were assembled, rigged and then towed behind transports to the airfields.

Security was paramount and Student went to considerable lengths to disguise the focus of the movements. His men were not permitted to advertise their presence in any way and had to disguise their appearance. The gliders and other potential giveaways were artfully concealed. All these precautions were, of course, rendered void by the operation against Corinth which blazoned the presence of airborne troops in the Greek theatre in the most dramatic manner. Student raged but there was little he could do to remedy the breach.

Worse was to follow. He had naturally counted on being allotted 22nd Air Landing Division, the obvious choice given their role in the earlier operations. OKW proved obstinate, however, citing transport difficulties as the reason why this formation should remain in Romania where it was engaged in the undemanding role of guarding the oilfields. As an alternative he was offered the 5th *Gebirge* (Mountain) Division, part of List's 12th Army and commanded by General Julius Ringel.

An Austrian by birth Ringel, as well as being a confirmed Nazi, was a soldier of the old school who had little respect for Student's more revolutionary ideas. His own were more pragmatic, his celebrated dictum 'sweat saves blood' summed up his methodical and cautious approach.[11] Liked and respected by his elite alpine troops, he was not the partner Student would have wished for. Lohr, commanding *Luftflotte 4*, appeared to favour his fellow Austrian and it was Lohr whom Goering had appointed as overall commander of Mercury.

As elements of his forces were already being moved to conform with the demands of Barbarossa, Goering may have felt that Lohr would ensure the attack on Crete was undertaken within the tight time limits the greater offensive permitted. General Korten, Lohr's chief of staff, was an intimate of the *Reichsmarschall* but no ally of Student, tending to the prevailing view that Crete was, at best, a distraction.

Even more trying, *Fliegerkorps VII* was commanded by an officer of equal rank to Student, Wolfram von Richthofen, a relative of the legendary Red Baron. A highly respected officer, who had pioneered and practised the aerial tactics of blitzkrieg, he was not easy to work with and Student had no direct authority over him.[12]

There were immediate differences in the manner of planning the assault. Student favoured a classic vertical envelopment, landing troops on top of each key objective which had been identified as the three air strips on the western section of the north coast – at Heraklion, Rethymnon and Maleme, together with Chania and the naval anchorage at Souda Bay. By attacking all at once Student would force the defenders to disperse their forces to a degree which would frustrate any determined counter-attack.

Lohr was sceptical; he favoured a more conventional approach, concentrating the drop around Maleme, the most westerly of the air strips which could be strongly held as a base for ferrying in reinforcements. Not only was Maleme the closest to German air bases in Greece, by concentrating the assault here, the air cover provided by von Richthofen's planes could be maximised.

This was the last thing Student wanted, a single blow that would enable the enemy to mass his reserves unimpeded by ground forces and deliver a strong counter punch, which would ensure the battle degenerated into an infantry slogging match. Adopting such an approach was to throw away the inestimable advantages of vertical envelopment. Richthofen, however, was adamant that his transport planes were insufficient to manage four drops in a single armada and that his fighters and bombers could not provide adequate air cover.

In the event the dispute was referred back to Goering and Jeschonnek returned to Athens with a compromise which, though it appeared sound on paper, really satisfied neither side of the argument. It was proposed that Student's vision be accomplished in two separate lifts; the western end, Maleme and Chania, would be attacked in the morning while the targets to the east, Rethymnon and Heraklion, would be left till the afternoon.

The disadvantages of the compromise were that the turnaround of aircraft would require perfect timing and the afternoon assaults would be delivered against adversaries already alerted. Surprise in the east was therefore lost and the scheme neither achieved Student's single irresistible blow nor Lohr's concentration. Student, remembering Holland, was aware how horribly vulnerable parachutists were when, immediately on dropping, they were confronted by an alert and determined foe.

The initial airborne assaults were to be delivered by the paratroops with Ringel's mountain division airlifted to whichever location appeared the most favourable. Student left the necessary training of the infantry to his subordinate Colonel Ramcke.[13] Despite being a highly competent officer, Ramcke's presence did nothing to allay Ringel's mounting concerns. He later wrote that when he first saw the orders he decided the mission verged on the suicidal.

It had originally been decided that, once the airborne bridgehead was established, some armour and heavy guns could be transported by sea. Student did not favour this, certainly not in the early stages, he preferred to wait until the harbours were secure and the British minefields dealt with. Hitler, however, was adamant. He remained convinced of the need for early reinforcement by sea, an insurance against initial failure or delays from the air.

Putting a suitable flotilla together, however, was no mean accomplishment. The German Navy had no effective presence and warships would have to be provided by the Italians, themselves badly mauled in the sea fight off Cape Matapan on 28 March.[14] Transports would have to be cobbled together from requisitioned caiques, small steamers, whatever could be found, enough for two flotillas, each of which would be responsible for shipping a battalion of alpine soldiers, together with some heavy guns and equipment. Parachutists would man light flak guns which, when later unloaded, could be incorporated into perimeter defences.

If Ringel's Alpine elite regarded the idea of going to war by Ju52 with trepidation, they were even less enthusiastic about taking to the water, particularly in a scratch built and ramshackle fleet, uncertain allies for protection and the might of the British Royal Navy against them. Most had never been to sea before and were dreading the prospect – their fears were to prove entirely justified.

From the outset the logistical difficulties loomed large, the vast and diverse range of matériel needed to sustain an airborne offensive had to be sourced, obtained and then transported, all in the teeth of a military administration wholly focused on another and far larger operation to the east. The tortuous route involved railways to the Romanian Black Sea Coast and then a time-consuming transfer to ships for onward transmission through the fateful narrows of the Dardanelles, scene of history's first great amphibious invasion – Troy – and, more recently, the disastrous Gallipoli offensive of 1915.

Fresh water and fuel were in equally short supply; the need for the former was solved by the acquisition of a bottling plant in Athens but getting in the vast stock of aviation fuel proved a quartermaster's nightmare. Some 24,000 cubic metres was required and the supply operation necessitated a major feat of engineering to clear the remains of the wrecked bridge which had collapsed into the Corinth Canal at the end of the previous airborne battle.

The available airfields were primitive in the extreme and the vast clouds of dust kicked up by the planes taxiing for landing or take-off became a major hazard; spraying the runways with water turned the whole lot into mud. Much of *Luftflotte* 4's maintenance and base personnel were already being siphoned off to meet the relentless countdown to Barbarossa; soldiers and drafted POWs had to be used to plug the gaps.

Richthofen's squadrons naturally took priority and gobbled up the best of what was available. Student was reduced to trying to utilise a dried up lake at Topolia, fifty miles from Athens.

By 14 May, however, Richthofen's aircraft were hammering the RAF bases on Crete and pulverising shipping in Souda Bay. By the 19th he was able to report he believed any threat of interference from British planes had been effectively neutralised. Even as the Stukas were disgorging their deadly loads over the island, Student summoned his senior officers to the impressively opulent setting of the Hotel Grand Bretagne for a detailed briefing.

The General was not interested in luxury, however, and any lingering doubts the battalion and regimental commanders might still have entertained about their intended target were soon dissipated. There was, as Heydte observed, a large scale map of Crete on the wall, a fairly significant clue.

Student, making the best of the compromise ordered by the Luftwaffe, had accepted that two waves of attackers would be required. The first lift would drop on Maleme and around Chania. With the airfields secured, Ringel's mountain division would be air landed the following day. Later, in the afternoon of the first day, Rethymnon and Heraklion would be attacked.

The attacking force would be split into three battle groups and *Gruppe West*, under Major General Eugen Meindl would comprise the entirety of *Luftlande Sturmregiment* (less two companies of glider borne troops who would participate in operations in the centre). Their prime objective was to secure Maleme Airstrip and the ground as far west as Kastelli. The drop would commence at 7.15 a.m. (German time) and Meindl could, in the afternoon, expect to be reinforced by the first flotilla bringing in a support battalion from Ringel's alpine division.

All through the briefing Student stressed the paramount importance of securing the air strips; success meant the difference between victory and defeat, life and death or capture. Once his objectives were secured Meindl was to advance rapidly eastwards to link up with the assault on Chania.

Lieutenant General Wilhelm Sussman would lead the first wave of *Gruppe Mitte* which would seek to land in the area known as Prison Valley[15] and then attack toward the coast and secure Chania and Souda. Sussman would jump with the divisional HQ, the two detached glider companies, and Major Richard Heidrich's 3rd *Fallschirmjäger* Regiment, together with various support units. His prime objective was to take and hold the island capital and neutralise the British command structure, thought to be located there.

The second wave would be under Major Alfred Sturm with 1st and 3rd Battalions of 2nd *Fallschirmjäger*. They were tasked with securing Rethymnon. Heraklion, the other main target of the second wave, was to be assaulted by *Gruppe Ost* under Major Bruno Brauer. He would command 1st *Fallschirmjäger* with 2nd Battalion of the 2 Regiment. Like Meindl at Maleme, *Gruppe Ost* could expect to be reinforced by sea during the course of the first afternoon's fighting.

Student, as the highly capable and conscientious officer that he was, had laid the ground for Mercury as well as he could despite the military objections, supply difficulties, breaches of security, his lack of overall authority and the general preoccupation with momentous preparations elsewhere. In one area particularly he was dangerously deficient and that was in the vital realm of intelligence.

Major Reinhardt, Student's intelligence officer, expressed the view that the total strength of the defenders would not exceed 5,000, most of them grouped toward the west, with only a handful, perhaps 400, at Heraklion. Even more fanciful was his assessment that the Cretans were only just waiting for an opportunity to rise and throw off the yoke of their supposed oppressors.

Ringel may have entertained private doubts but Student was anxious to minimise his subordinate's role in the attack. The Austrian was expected to wait until the second day when he and his alpine troops could be ferried onto the captured airstrips. Student, typical of his leadership style, intended to assume personal command once the bridgehead was established; this was, after all, his operation and the masterpiece must bear the stamp of its creator.

As for the officers and men, at this stage they had no reservations, they were the cream of the German armed forces and they had never tasted defeat; they had an absolute trust in their commander. As the battalion and regimental commanders briefed their men the following day, the *Fallschirmjäger*, clustering on the makeshift airfields or under the shade of olive groves which offered some relief from the relentless Aegean sun, would not have entertained any thought of failure.

Most would have felt a surge of enthusiasm and confidence; the endless, weary miles by road and rail, the backbreaking, sweating labour of manhandling oil drums and supplies over the dust-laden ground were, at last, at an end. The prospect of action loomed ...'you are the chosen ones of the German army. You will seek combat and train yourself to endure any manner of test.'

In the gathering twilight of 19 May the men prepared to emplane, taken by trucks to the airfields. They sweated in the stifling, dust laden air, weighed down with uniform and kit – the paratroopers were still wearing their standard field pattern, intended for colder climates and had not been issued with tropical gear.

Von der Heydte recalled the urgency and confusion of that night:

We were greeted by the ear splitting roar of a hundred and twenty ... transports as they tested their engines ... Through clouds of dust we could see red glowing sparks flaring from the exhausts of the machines, and only by this light was it possible to discern the silhouettes of our men. Flashing the pale green beams of their torches ... the officers and NCO's of my battalion tried their best to make themselves heard above the thundering of the engines ... It was a few minutes after 4 a.m. when my aircraft taxied onto the runway. The first light of dawn scarcely penetrated the red dust raised by the machines during the night, which hung like a dense fog over the airfield.[16]

Operation Mercury was now under way and the die was effectively cast; as Churchill later wrote:

> The story of Souda Bay is sad ... how far short was the action taken by the Middle East Command of what was ordered and what we all desired ... it remains astonishing to me that we should have failed to make Souda Bay the amphibious citadel of which all Crete was the fortress. Everything was understood and agreed, and much was done; but all was half scale effort. We were presently to pay heavily for our shortcomings.[17]

The Prime Minister wrote with a politician's fine eye for hindsight and selective memory and with a neat gloss over what was ordered and what was achievable in practice. The debacle of Greece had emasculated Wavell; under pressure from the Axis in the desert he simply did not have the resources to turn Crete into the island fortress of Churchill's imagination.

With the Luftwaffe completely in control of the skies it was questionable from the outset if Crete could be held. The Navy would be horribly exposed without air cover and Cunningham was already fully stretched, the consequences for the bare margin of British maritime supremacy in the Eastern Mediterranean could so easily prove fatal.

As the evacuation from the mainland progressed during the final days of April some 20,000 additional troops reached the island.[18] Most were exhausted and demoralised, without weapons or kit, frequently disorganised without their own officers and NCOs. In some cases they were little more than a mob, their already frayed nerves further sapped by incessant and costly harassment from the air.

Souda Bay was like a vision of the inferno, Stukas and Dorniers pounded the docks at will and without respite; one of the two tankers hit by German bombs, the 10,000 ton *Eleanora Maersk* was still burning fiercely. A dense, pungent pall of greasy smoke blanketing the bay, garnished with the bloated corpses of lost sailors. These were not sights likely to restore anyone's confidence.

Once they finally got ashore, mostly being transferred by lighters, the troops found that there was nothing and no one to receive them; no barracks, no tents, not much in the way of supplies, they were obliged to bivouac as best they could among the olive groves. Many took to foraging. Hunger, exhaustion and the heady Greek wines proved a dangerous mix and there was some disorder before the situation could be restored.

If there was a lack of military police, junior officers and NCOs, there was no shortage of generals; in fact there was something of an embarrassment. General Weston, then in command of the island's defences, found himself outranked by 'Jumbo' Wilson and matched by Mackay and Freyberg. Wavell compounded the problem by trying to treat the two senior commanders as equals and addressing communications to both. Wilson's days were already numbered and his pessimistic, if by no means inaccurate, summary of the island's vulnerability probably proved the final nail.

The plain fact was that nobody had really considered what might happen if Greece fell and the island suddenly became the new front line. The task facing the Allied commanders in April 1941 was therefore a formidable one – the long exposed ribbon of Crete, ill served by roads and with only the three airstrips, had to be protected against a conquering foe with complete air superiority.

As far back as October 1940 when General Metaxas had first bruited the question of the British Guarantee, Middle East Command had preferred to limit its involvement to an occupation of Crete. Papagos was later promised support from the air and sea with a full brigade to be dispatched to bolster the local conscripts defending the (then) only air base at Heraklion. The vulnerability of this single strip to parachute attack was considered although, in the event, Operation Marita included no plans for an attack on Crete.

Once the Italians had begun their abortive invasion Wavell moved quickly to secure the island. A small detachment was sent out directly and 2nd Battalion Yorks and Lancaster, deployed in Operation Action soon followed. This spearhead was to be bolstered by HQ 14 Infantry Brigade, 2nd Black Watch, one heavy and one light anti-aircraft (AA) units and a support company; a total commitment of 2,500 soldiers.[19] The initial reports concluded that Souda and Heraklion were defensible, the anchorage permitted the offloading of heavy weapons and the local populace, both military and civil, were enthusiastic.

The Cretan troops available comprised some 7,000 conscripts from the 5th Cretan Division, backed by a slightly larger number of reservists and 1,000 paramilitary police. The mountain men were natural fighters and excellent soldiers though, as was remarked, 'inclined to individualism rather than the team spirit'.[20] The Gendarmerie had been raised to act as watchdogs and repressors –

there to curb the tendencies of the local republicans.

Brigadier O.H. Tidbury, the first garrison commander, thus moved to deploy his modest assets in defence, primarily of the airstrip at Heraklion, while the individual Tommies soon found their Cretan allies possessed a splendid concept of generous hospitality, cementing a warm regard that would survive invasion, defeat and occupation. At the same time egg and chips made its first appearance on local menus. If the Cretans needed any rallying call it was provided by their clergy, who were united in their steadfast loathing of the Axis, theirs a powerful voice in so devout a society.

Italian prisoners of war, many in wretched condition, were another Allied import, in their sullen thousands – little by way of elaborate security was required, their justified terror of the undisguised fury of the locals was more than sufficient!

As the fighting in the harsh northern mountains persisted the Cretan Division was drawn off and fed into the fight. Many of the islanders' donkeys were also conscripted. Although understandably proud of their compatriots' distinguished role in the war, many locals were suspicious that the national government was using the Italians as a ruse to strip the place of able bodied men. Throughout, the island was to be starved of arms and ammunition, to the great detriment of the resistance, so deep did the government's fears stick.

Despite these encouraging factors, Churchill's vision of an impregnable island fortress was destined to be stillborn. The lack of air cover was a crippling limitation and Wavell, at the outset, saw the place as little more than a naval bulwark. The commander-in-chief was understandably preoccupied with operations in the desert and then, latterly, on the Greek mainland.

Churchill blindly maintained that Wavell had men to spare. He was obsessed by the view that the ratio of fighting troops to support personnel was disproportionate, failing to understand that the logistical demands of armoured warfare in desert conditions require a huge logistical effort. By November 1940, however, it all began to seem academic. Middle East Command was concentrating on the forthcoming desert offensive and the Greeks were more than holding their own against the crumbling Italian offensive. Both Wavell and Cunningham, fearful of overextending the slender resources, agreed that Crete should stay as a backwater.

Although plans were made to stiffen the garrison in the event Greece fell, and a Commando group was deployed on the tip of the east coast both to guard against Italian raids from Rhodes and mount a few of their own, the position largely remained static. The locally raised forces were now only a few thousand strong with a distinct paucity of serviceable arms. Tidbury had, from the outset, envisaged that the likely threat would come from the air and had therefore concentrated on digging in around Souda, constantly hampered by a lack of engineering tools and equipment. With so many young men away in distant Epirus, local labour was equally as scarce.

Tidbury was saved from further anxiety by his transfer; the first, but by no means the last, commander to arrive and depart in haste. In the following six months several officers were appointed and replaced in rapid succession. Major-General Gambier Perry followed Tidbury but was soon supplanted by Brigadier Galloway with a diminished brief to safeguard Souda Bay only.

His replacement, Brigadier Chappel, was the first to seriously question what he was supposed to be about. Was he supposed to formulate a comprehensive plan for the defence of the island? As the Axis was, by this point, poised to invade the mainland, there was a greater urgency to these considerations and Chappel deployed his existing, slender, forces as best he could with a series of anti para- chutist exercises included in training.

Further confusion ensued. The demands of the looming Greek campaign prompted Wavell to consider consolidating the defences at Souda to provide Cunningham with a secure anchorage, at that time, beyond the German bombers based in Bulgaria. On 2 April Major General E.C. Weston commanding MNDBO was directed to take charge of Souda. The vulnerability of this key anchorage had been amply demonstrated a few days earlier when a daring raid by Italian light craft had decisively crippled the cruiser HMS *York*. Weston was sent immediately to Crete while his troops travelled more circuitous- ly through Haifa but the matter of whether he or Chappel was to hold overall command remained unclear.

Before the position was clarified in Weston's favour on 27 April the general had already produced a detailed assessment. He submitted this on the 15th by which time the likelihood of a defeat on the mainland loomed large. He believed that the island could be attacked by both airborne and seaborne forces and that these attacks could occur in the east, the west, or at Rethymnon.

A greater measure of defensive capability was therefore required, to be dispersed to meet these numerous threats though he did feel the eastern end could be entrusted to Greek forces, provided sufficient were available. He stressed that more air strips, together with an adequate complement of fighters and bombers were an essential element in a successful defensive strategy. This in turn raised the necessity of greater AA cover.

These conclusions were substantively endorsed by the Middle East Joint Planning Staff on 21 April, by which time it was abundantly clear the fight for Greece had been lost. They decided that, in order to bring the island up to a proper state of readiness, three brigades would be needed with adequate air and anti-aircraft support.

Like Weston, they saw a major threat from the sea supported by paratroops – no one yet had considered the potential for vertical envelopment only, no one except General Kurt Student. This question of continued coastal defence was to become a kind of mantra which would bind the local commanders in a pernicious vice while the main struggle was being fought out around the airfields.

It was proposed that units evacuated from Greece would not be deployed as part of the defence plan, they would be withdrawn to Egypt to recuperate and refit. These two sets of recommendations were sound but circumstances and the increasing demands of a sector under threat from all sides, from Rommel in the desert to insurgents in Iraq, robbed Wavell of the opportunity to react accordingly.

At this point even Churchill did not appear to perceive Crete as a priority; his directive to Wavell of 18 April contains only a cursory and dismissive reference:

> Crete will at first only be a receptacle of whatever we can get there from Greece. Its fuller defence must be organised later. In the meanwhile all forces there must protect themselves from air bombing by dispersion and use their bayonets against parachutists or airborne intruders if any. Subject to the above general remarks victory in Libya counts first, evacuation of troops from Greece second. Tobruk shipping unless indispensable to victory, must be fitted as convenient. Iraq can be ignored and Crete worked up later.[21]

The suggestion that the defenders engaged airborne troops with their bayonets reflects the Prime Minister's romantic view of war – deeds of derring-do which contrast starkly with his acute grasp of political realities. His failure to realise that Wavell's ability to wage war successfully in the desert theatre without adequate support personnel stems from the same view, a harking back perhaps to the more chivalrous days of Victorian colonial wars.

Adequate air defence had been a key tenet of both Weston's and the Joint Planning Staff's reports. And here was the rub. The RAF was virtually bankrupt in terms of available aircraft and personnel. The demands of the desert war, Greece and the agonisingly slow re-supply left Air Chief Marshal Longmore with an empty cupboard. Greece had drained his slender resources for no appreciable gain.

During the winter of 1940/1941 activity on Crete had been limited to the strictly non operational – the construction of a series of new airstrips. By April only those at Rethymnon and Maleme, the latter in the extreme west, had been completed. Plans for additional facilities even further west at Kastelli, also at Pedalia and Messara Plain had not come to fruition, largely due to a dearth of labour and materials.

Even once completed these air fields were bare strips only, apart from Heraklion none had pens for the aircraft to be deployed, assuming any were to be deployed. The uncertain status of Crete in the overall strategic situation, the lack of detailed planning and inter-services liaison was to bear bitter fruit. Radar was not available until April 1941 when one centre was set up at Maleme and a second was under way at Heraklion. Communications to the fire control room at Chania, which was to coordinate AA batteries, was via a single, vulnerable telephone line.

When the German threat to Greece first materialised there was a suggestion that the RAF should base its response in Crete. This idea, promulgated by Air Chief Marshal Sir Charles Portal, whist eminently sound, was abandoned in favour of a disastrous forward strategy, placing the available squadrons on the mainland. Once there the hopelessly outnumbered squadrons were decimated, mainly while still on the ground. Crete was defended only by elements of the Fleet Air Arm, equipped with a motley of obsolete and frequently unserviceable aircraft.

By the conclusion of the Greek debacle the RAF had lost 209 planes, seventy-two destroyed in combat, fifty-five shot up on the ground and the remainder wrecked during the evacuation.[22] The rate of loss outstripped the supply of replacements. Longmore's frequent appeals for increased supply won him few admirers in the War Cabinet. The politicians, removed from the day to day realities, appeared unable or unwilling to separate 'paper' planes from those able to fly! At the start of May he was recalled and Tedder, his deputy, was appointed. As ever, shooting the messenger did not solve the problem.[23]

By the middle of April Wing Commander G. R. Breamish had been sent to Crete to take charge of the air defences. His forces, at that point, comprised less than twenty personnel, though this meagre complement was boosted by the arrival of No. 30 Squadron with eighteen Blenheim bombers together with the remains of three fighter squadrons. After their dire losses on the mainland these, combined, could only show sixteen Hurricanes and six outmoded Gladiator biplanes; none in good repair.

Such depleted resources were insufficient to daunt the Luftwaffe and the continuing demands of North Africa meant that there was scant prospect these could be significantly boosted. Superiority in the air would therefore continue to lie with the Axis, effectively cancelling the strength of the Allies at sea and raising grave questions over whether the island could in fact be successfully defended at all.

The nature of the island's geography, the long exposed strand of the north coast, linked by only a single road, the airfields, basic and largely undefended and with no facilities for dispersal, would offer a tempting target for the Luftwaffe; likely a repeat of the experience on the mainland. Once the Axis had constructed its own strips in the Peloponnese the island would be in range of the dreaded Me109s with their lethal capacity for low level strafing.

Blame for the eventual defeat was shunted onto Wavell, tidily ignoring the immense practical difficulties confronting GHQ Cairo and disregarding the natural obstacles created by the topography. To establish a viable route to the south coast (the road to Sphakia was incomplete) would have been a major engineering feat, impossible in the time available and, even if completed, suicidal for convoys in the face of German air power. It can also be said that it was the decision to engage the enemy on the mainland that diverted attention away from Crete. Had the concept of the island fortress been actioned at the start of 1941 then history might have taken a different course.

Greece had eaten into Wavell's supplies, meagre at best, to an alarming degree:

The loss of men was ... mercifully lighter than it might have been: 2000 had been killed or wounded and 4000 made prisoner out of 58000 troops sent to Greece. But the loss of materiel was disastrous: 104 tanks, 40 anti aircraft guns, 193 field guns, 1812 machine guns, about 8000 transport vehicles, most of the signals equipment, inestimable quantities of stores and 209 aircraft... [24]

With the mainland lost Churchill moved the island up to a higher slot in the list of priorities – the question was whether to seek to hold Crete or withdraw all units to North Africa. The Prime Minister had no doubts:

> Crete must be held [he instructed Wavell on 17 April] ... and you should provide for this in the re-distribution of your forces. It is important that strong elements of Greek Army should establish themselves in Crete, together with King and Government. . we shall aid and maintain defence of Crete to the utmost.[25]

The decision was taken and he would not brook any contrary opinion; Cunningham was rebuked for expressing reservations. Wilson then sent his pessimistic but realistic assessment but the die was firmly cast. With ULTRA intelligence now playing a part,[26] Churchill launched fully into bulldog mode: 'It seems from our information that a heavy airborne attack by German troops and bombers will soon be made on Crete ... It ought to be a fine opportunity for killing parachute troops. The island must be stubbornly defended.' [27] Moreover, the PM knew just the man to exploit this heaven-sent opportunity for slaughtering the enemy.

General Bernard Freyberg was indeed a Herculean figure – physically imposing and utterly fearless, he had served with considerable distinction during the Great War and had won the Victoria Cross. Churchill regarded him with some reverence. Although born in New Zealand, most of his life and career had been in Britain. When the Prime Minister, meeting the General in the 1920s, begged him to 'strip his sleeve and show his scars' , he counted no less than twenty-seven old wounds.

It was to this officer, the doyen of fighting soldiers, that Wavell turned at the end of April. He had lost confidence in Wilson who clearly had little confidence of his ability to hold the island. Flying to Maleme on 30 April, the commander-in-chief interviewed Wilson and informed him of his decision to post him to the Levant. Having disposed of one commander he then spoke privately to Freyberg who was nearby. After congratulating the General on the performance of his Kiwis in Greece he dropped the apparent bombshell that Freyberg was now to command in Crete.

Perhaps any general, on being entrusted with so difficult an enterprise, would baulk. Freyberg would naturally do his duty and obey orders but he was, of course, subject to separate political constraints in that he reported directly to the home government. Wavell was not to be deflected; he intimated that the decision emanated from Downing Street and could not be gainsaid. Freyberg, loyal and naturally chivalrous, could do no other than accept. Possibly, even at the outset, he felt his command to be a poisoned chalice.

The situation on the island was far from propitious. Since the British presence had first been established in late 1940 no clear vision for the defence had been realised. It was a case of 'muddling through' and this lack of strategic overview and the absence of a coherent plan was palpably obvious.

SOE had sent Peter Wilkinson to observe the state of preparedness and his conclusions were far from encouraging. He reported on a state of 'complete inertia' – a total lack of 'elementary precautions'. He commented, tellingly, on the lack of any decent north/south road, despite the fact that there was only some four miles left to complete and the garrison had had six months in which do the job.

That the lack of a major arterial road to the south coast was a considerable weakness, was scarcely a revelation. The Axis's success in Greece had placed the great harbours of the north coast within reach of the German bombers. For any re-supply to be effected from Egypt by sea would expose the British ships to the gauntlet of air attack, at a time when the RAF was so seriously depleted. Souda Bay had become a maritime graveyard with over 50,000 tons of Allied shipping already lost. Even with the inestimable boon of hindsight it is possible to see that had Sphakia been turned into a viable small supply harbour, with the steep mountain road to Askifou completed to a reasonable standard, the risk to the ships would have been considerably diminished.

Most disturbing was the lack of air cover. An observer wrote of the doomed flight of the final Hurricane to take off from Maleme, instantly swallowed by a horde of marauding Messerschmitts. Desperate as the odds were the situation had been considerably exacerbated by the poor siting of AA batteries and the failure to construct fighter pens and smaller, satellite airstrips under the sheltering lee of the high hills.

In his report Wilkinson did not spare the Air Force: '...the attitude of the RAF beggars description'. Unconvinced by 'excuses', he draws a most unfavourable comparison with the Luftwaffe, citing their apparent ability to carve out temporary airfields within hours of their arrival. Most tellingly he points out that, whilst prior to the

Greek debacle Crete may have been an inconsequential backwater, the evacuation from the mainland put the island in the strategic forefront. Lastly Wilkinson castigated the Navy for the poor state of preparation at Souda Bay, citing the lack of any foam firefighting apparatus.

Upon assessing the burden of his command, Freyberg sent an urgent signal to Wavell wherein he complained the force he had available was inadequate and that he needed support both at sea and from the air. He pointed out that much of his troops' heavy equipment, particularly artillery, had been abandoned in the course of the Greek fiasco. Even entrenching tools were in pitifully short supply, as indeed was just about everything else. At the same time he wrote in a very similar vein to the home government.

The C.-in-C., having conferred with Admiral Cunningham, responded in a positive tone, giving assurance that the Navy would not let the defenders down – this was in spite of the Admiral's misgivings that, at such short notice, the island could be evacuated. Wavell, throughout, was not convinced of German intentions by sea. Neither he nor Cunningham believed the Axis could amass sufficient vessels for such an undertaking.

Churchill, in London, also perceived the greatest threat lay more toward the skies than the clear, blue waters of the Mediterranean. When the Prime Minister wrote to his counterpart in New Zealand he emphasised the nature of the airborne threat, at the same time persisting in the Homeric view that the able-bodied Kiwis would relish a straight fight with an enemy, who lacked the decisive support of tanks and heavy guns, 'on which he so largely relies'.

Part of the difficulty lay in Freyberg's own mercurial temperament. He certainly lacked faith in his own capabilities and yet was motivated by an admirable sense of duty. His mood swung from almost feverish optimism to deep despair. This may, at least in part, explain[28] that when he wrote to Churchill on 5 May his tone was far more bullish: '...cannot understand nervousness; am not in the least anxious about airborne attack; have made my dispositions and feel can cope adequately with the troops at my disposal.'[29]

Was it simply the case that, having been entrusted with his mission by Churchill, Freyberg simply felt he could not pass the challenge. The effect of the Prime Minister's charisma should not be overlooked. If the politician admired the soldier then the soldier was bound to do his utmost to conform to the image he had inspired. There were few men of whom Churchill would write:

At the outset of the War no man was more fitted to command the New Zealand Division, for which he was eagerly chosen. In September 1940, I had toyed with the idea of giving him a far greater scope ... Freyberg is so made that he will fight for King and country with an unconquerable heart anywhere he is ordered, and with whatever forces he is given by superior author- ities, and he imparts his own invincible firmness of mind to all around him.[30]

Stirring stuff but it betrays the great man's romantic weakness of assuming that a strong heart and a just cause can overcome all odds. Modern warfare is not that accommodating; it is, in part, an indus- trial process, weight of men, matériel, supply and above all air power will generally decide the issue. Both men fell into the same trap, Churchill elevated the hero and the hero had to conform, whatever his professional misgivings.

Supply was an immediate problem. German bombing had made Souda too hot a landfall in daylight hours and ships had to be unloaded in darkness. Damaged ships lay lifeless in the water and their precious cargoes, needed to supply such a swollen garrison, had to be manhandled. In the first three weeks of May, immediately prior to the attack, some 27,000 tons of munitions were embarked for Crete but only a pitiful percentage, some 3,000 tons, reached the dockside.

The situation did begin to improve somewhat when Major Torr took over responsibility, backed by contingents of Australian volun- teers, from engineering units and 2/2nd Field Regiment.[31] The increased energy and efficiency had an effect, some Bren carriers, apparently lost on a half submerged wreck, were ingeniously salvaged and made operational. A frantic nightly run by fast destroy- ers, speeding into Souda, unloading and dashing back to Alexandria under the sheltering blanket of darkness, also eased the crisis.

Additional arms in the shape of a motley collection of French and captured Italian 75-mm and 100-mm guns, a battery of mountain guns, 3.7-in howitzers, together with assorted armour, sixteen light and half a dozen infantry tanks were obtained.[32] The MNBDO were a considerable addition in themselves, apart from the 2,200 marines, they were fully equipped with light and heavy AA guns, searchlights and some formidable 4-in naval guns. A brace of fresh contingents which, like MNBDO, had not been exposed to the debacle in Greece,

2nd Leicester and 2nd Argyll and Sutherland Highlanders, came ashore.

Admiral Cunningham, whatever his reservations about the safety of his fleet exposed in open waters to an enemy with full control of the skies, had pulled together two 'heavy' flotillas and seven 'light'. The battlecruisers with their massive 15-in guns stalked the western approaches to deter any intervention by the Italian Navy, while the squadrons prowled the coast ready to pounce on any invader. In total the fleet comprised four battleships, nineteen cruisers and forty-three destroyers. Such a concentration of sea power virtually doomed any attempted landing to certain destruction. Churchill, writing to Mr. Fraser in New Zealand felt the odds were now even:

> The Navy will certainly do their utmost to prevent a seaborne attack, and it is unlikely to succeed to any large scale. So far as airborne attack is concerned, this ought to suit the New Zealanders down to the ground, for they will be able to come to close quarters, man to man, with the enemy who will not have the advantage of tanks and artillery, on which he so largely relies. Should the enemy get a landing in Crete that will be the beginning, and not the end, of embarrassments for him. The island is mountainous and wooded, giving particular scope to the qualities of your troops.[33]

This schoolboy romanticism has been dismissed as patronising but this is probably an injustice; it more reflects the Prime Minister's enthusiasm for a good clean fight, chivalric warfare with cold steel to the fore. In the event the New Zealanders more than justified his confidence in their martial spirit.

Freyberg was faced with the reverse of the tactical dilemma which had earlier confronted Student. The coastline was long, the airstrips widely separated. Should he therefore protect each of his strategic assets in force, thus spreading his resources and exposing them to defeat in detail, or should he rather thin out the defenders and build up a strong reserve, available to be rushed to the contact in sufficient strength to re-take any bridgehead the enemy might win?

Again, like his German opponent he opted for a compromise, splitting his troops into three principal contingents, each charged with the security of a vital sector but leaving the final deployment in each case to local commanders. Like their General many of the officers involved, brave and dedicated soldiers, were veterans of trench warfare and commenced 'digging in' and wiring their, pre-dominantly linear, positions. This gave Freyberg great comfort and he wrote in a more confident frame to Wavell after his tour of inspection on 13/14 May.

In the vulnerable west of the island, the Maleme/Galatos sector, he deployed the New Zealand Division; 4 Brigade under Brigadier Inglis comprising 18th, 19th and 20th (NZ) battalions and the 5th commanded by Hargest – 21st, 22nd, 23rd (NZ) and 28th (Maori) battalions. Brigadier Howard Kippenberger led the newly created 10 Brigade (NZ Divisional Cavalry Detachment and Composite Battalion).

Weston remained in command of his MNBDO dug in around Souda and supported by a pair of composite Australian battalions, together with 2/2nd Field Artillery (deployed as infantry). Freyburg's 'Creforce' HQ was near Chania, the administrative capital and the Force Reserve – 1st Royal Welch Fusiliers, (14 Brigade), 1st Ranger Battalion (9th Battalion KRRC) and the Northumberland Hussars (Noodles) was nearby.

Moving eastwards along the long ribbon of the north coast, the Rethymnon/Georgioupolis sector was held by Brigadier Vasey's 19 Australian Brigade. This comprised 2/1st, 2/7th, 2/8th and 2/11th Battalions, all infantry; three batteries of guns from 2/3rd Field Regiment together with units of field engineers and machine gunners.

The 14 Brigade, commanded by Chappel, was deployed around Heraklion and his forces included the 2nd Black Watch, 1st Argyll and Sutherland Highlanders, and the Australian 2/4th Infantry Battalion, (part of 19 Brigade).

This left only the Greek contingents. There was a tendency to regard these as distinctly second rate formations; they were badly armed and equipped but they were to show that there was nothing lacking in their fighting spirit. They, together with ad hoc groups of local irregulars, would make a significant contribution to the defence. The Germans would be shocked and on numerous occasions discomfited by the ferocity of the local response, being accustomed to a more servile reception. It was easy to forget that the Cretans had a long and proud tradition of offering fierce and unbending resistance to the invader.

General Freyberg, whatever his other failings, was quick to appreciate the fighting qualities of the Cretans; theirs was a spirit to which this lion of a man could respond – simple, courageous, fiercely independent and resolute. He realised that local bands of guerrillas or andartes could harass and decimate an invader; their lack of training and arms amply compensated by centuries of resistance, and intimate knowledge of the difficult terrain.

There was a tendency amongst both British and dominion officers to write off all the Greek units as equally unreliable. This was grossly unjust. Many were indeed crammed with raw recruits and not all would perform well. Kippenberger was particularly dismissive and, in some units, morale was undoubtedly low. The erratic provision of arms, which was miserly and random, was hardly calculated to stiffen anyone's resolve. The men drew their rifles from a central depot housed in Chania where they faced an eclectic choice of mainly outdated weapons with, at best, a few rounds apiece, frequently of the wrong calibre!

Colonel, later Brigadier, Guy Salisbury-Jones was given the job of liaison officer with the Greeks, and their eight battalions comprised a total of some 9,000 effectives. Of these Freyberg now deployed the 1, 6 and 8 Regiments in the far, western sector, the 2 Regiment in Souda/Chania, with the 4th and 5th together with the paramilitary gendarmes in the Retymnon/Georgioupolis sector. The remainder, 3 and 7 Regiments and a Garrison Battalion remained at Heraklion.

The New Zealand Division was under the command of Brigadier General Puttick who shared part of his sector with elements of the MNBDO under Weston, disgruntled at his replacement by Freyberg, but who had responsibility for some AA and coastal defence emplacements. Weston's semi-independent fiefdom did not make for smooth coordination. Force Reserve was not under Freyberg's direct command but was to be 'administered by sector commanders' – a rather woolly arrangement that was to have serious consequences.

Defence of the vital Maleme sector was designated by Puttick to Hargest's 5 Brigade; Kippenberger's 10th (less the 20th Battalion) was guarding Galatos and Prison Valley. The 20th, based on the fringes of Chania, was kept in hand as a distinct divisional reserve, not to be deployed without Puttick's acquiescence. Brigadier Hargest's dispositions around Maleme itself were to be crucial to the outcome of the forthcoming battle.

The 21st and 23rd were deployed on the high ground around Kondomari with his HQ further east at Platanias, some distance from the vital airstrip. A deep gully, the Sfakoriako, divided these units wired in positions from Maleme. His engineering battalion straddled the coast road by the bridge at Modhion and the crack

Maoris further back at Platanias.

The vital bastion of Hill 107, which overlooked the airfield, was to be defended by the 22nd battalion, commanded by Lieutenant Colonel Andrew VC, and itself dispersed to a degree that only a single rifle company occupied the summit. A major failing of the defensive enceinte was that no troops were stationed west of the dry Tavronitis riverbed and the iron bridge which spanned the arid watercourse remained unguarded. D Company's extreme right lay against the banks as did the slit trenches of C Company's No. 15 Platoon. The remaining platoons were dug in around the field itself. Worse, an improvised sprawl of tents and hutments, home to the RAF and other non-combatant personnel, crowded the line of fire.

This was not the only shortcoming of the defence of Maleme; the lighter AA guns, manned by marines, were controlled by Weston, the heavier AA weapons took their orders from the Gun Control Room in Chania. Though Andrew made representations to Puttick about the lack of deployment west of the Tavronitis nothing was done and the 'back door' remained ajar. Moves were considered to move some of the Greeks at Kastelli further east but all that was finally done was to detail a single section from 21st Battalion to occupy a single vantage as, to all intents and purposes, observers.

Air Marshal Longmore, having inspected the Souda Bay area, concluded it could be kept clear of German bombers by a full squadron of Hurricanes, with 100 per cent replacement rate and reserve of pilots. This may have been optimistic but Freyberg hoped he would at least receive this complement of fighters. He was to be disappointed as the debate raged at the very highest level over how the supply of available planes and pilots was to be doled out. The RAF sought to shift the burden onto the Navy; Cunningham hedged in response.

The final proposal was the worst of bad compromises. The island, it was felt, could not be adequately defended from the air given the heavy demands made by operations in the Western Desert but, and this was crucial, the air strips with their attendant personnel were to be kept at operational readiness. The idea was that, should the Germans invade from the sea, aircraft could then be dispatched from Egypt and based at these forward strips. For this reason the runways were not lighted.

By mid May the few available aircraft, stationed on the island, had been steadily whittled down by attrition. By the 19th Beamish had convinced a reluctant Freyberg that the battered survivors should be withdrawn from what was becoming a hopeless fight. This must have been a bitter moment; the fearful scenario he had envisaged at the outset had become a reality, his forces were without air cover, exposed to the full fury of the Luftwaffe.

There was, however, on Crete one officer whose preferred defence against parachutists was his trusty sword stick. John Pendlebury, a field archaeologist and Old Wykehamist in his mid-thirties, was one of those brilliant mavericks, in the vein of T. E Lawrence or Orde Wingate. He'd been curator of Arthur Evans museum collection at Knossos in the previous decade and knew the island and its people intimately. Military Intelligence had recruited him as early as 1938 to serve in what would become the Special Operations Executive (SOE); at that point Military Intelligence (Research) MI(R).

First dispatched to Greece in the wake of the German onslaught on the Low Countries in 1940, his credentials overcame the Greeks' suspicion of British clandestine operations on their soil; at this point Metaxas wished to avoid provoking the Axis. Pendlebury soon made his way to Crete where he had deep knowledge of the landscapes, gleaned in the course of his many walking trips around the island. In addition to his sword-cane he also sported a glass eye which he took to leaving on his desk when engaged in the field!

His official role was that of Vice Consul in Heraklion but he was soon abroad, organising the Palikari – such was his charisma that he was soon able to report he had established the basis for an intelligence and, if needed, resistance network. As MI(R) began its complex transformation into SOE, Pendlebury was somehow overlooked and continued, unfettered by official constraint, on his own initiative. One of his concerns was the reluctance of the government to arm the Cretans, still fearful of their republican sympathies.

Two of his intelligence colleagues, Terence Bruce Mitford and Jack Hamson, were sent in as support with a particular brief for potential sabotage operations against the Italian invaders on the mainland. However, with the Axis advance halting under way, Pendlebury was appointed as official liaison with the Greek army.

Having neatly circumvented the territorial restrictions imposed by demarcations within the intelligence organisations, he was able to set up a school for saboteurs on Souda Island. Nonetheless the severe restrictions placed on the team's remit by Cairo meant that by the time Greece had fallen and an invasion of Crete appeared imminent, frustratingly little had been achieved.

The mainland debacle added a sudden jolt of real urgency and the

SOE activity on Crete was geared up accordingly. It was now proposed to make the island a training ground and operational base for guerrilla activity throughout the Axis occupied Balkans. A new SOE HQ was established in a pleasant villa in Chania and Pendlebury's brief, to recruit local resistance groups, was given fresh impetus.

For a brief moment, the troops on Crete, despite the regular attentions of the Stukas, gained a respite. The island, largely untouched by the war, was lovely in the Mediterranean spring, the air heavy with the scent of thyme. Those early days in May were an opportunity to recover from the ordeal of Greece:

> I had a most heavenly bathe this evening with David Barnett. As I told you this is the most beautiful place and we found a lovely little sandy cove, surrounded by rocks, about three miles away. The water was crystal clear and just cool enough to be refreshing, with a pale blue tint. We sat on the rocks and dried in the evening sun, which doesn't burn you here. It wasn't like war at all.[34]

As expectation lay heavy in the scented island air and the troops, battered by their ordeal in Greece, recuperating in the glorious warmth of the Cretan spring, basked in the delicious shade of the olive groves or swam in the revitalising waters, General Freyberg spent his days touring the defences. The presence of this great, bluff bear of a man brought heart to many young soldiers. His very obvious concern for their welfare, his legendary valour and his curt injunction just to 'fix bayonets and go at them as hard as you can' were reassuring.

Some of his officers were a deal less sanguine, however. Their general's preoccupation was with resisting a seaborne threat, his obsession with the business of 'wiring in' – straight from 1918. This appeared to be the antithesis of rapid reaction and relentless counterattacks which was the accepted response to parachutists.

Freyberg had, in his operational orders, placed reliance on the force reserves. These were considerable but their correct deployment remained dependant on a number of factors – sound communications, speed and cohesion of response. The plain fact was that the reserves were scattered along the ribbon of coast without sufficient transport and, above all, with poor communications.

Here lay the nub of the problem. Many wireless sets had been lost in Greece, those which had been salvaged were few in number and, at best, indifferent in quality. Amazingly, Freyberg had not included radios in his 'urgent' list sent out on 7 May. Communications otherwise depended on field telephones, the wires strung precariously on poles along the length of the coast road.

Field telephones had proved totally inadequate in the previous war and were particularly vulnerable to casual interdiction by paratroops. Even the signal lamps had no batteries or were of the wrong voltage requirement for the fitful mains. This ad hoc and grossly inadequate system of communications was a major and telling weakness.

Due to strict adherence to Air Ministry requirements,the airstrips had not been mined or slighted and Freyberg's vision of a descent from the sea rather than the air continued to blind him to the threat, should the Germans prove able to snatch one of the aerodromes intact. On 16th he sent a final, pre-invasion signal to Wavell in Cairo, the tone upbeat and confident:

[I] have completed plan for the defence of Crete and have just returned from final tour of defences. I feel greatly encouraged by my visit. Everywhere all ranks are fit, and morale is high. All defences have been extended, and positions wired as much as possible. We have forty-five field guns placed, with adequate ammunition dumped. Two infantry tanks are at each aerodrome. Carriers and transport still being unloaded and delivered. 2nd Leicesters have arrived, and will make Heraklion stronger. I do not wish to be over-confident, but I feel that at least we will give excellent account of ourselves. With help of Royal Navy I trust Crete will be held.[35]

This then was the state of the island's defences on 19 May; time had now run out. Churchill was right to point out that the loss of Crete was indeed sad. It was to be, even on the most lenient of assessments, an avoidable defeat. Inertia, lack of organisation, lack of a coherent strategy for the defence, and the under use of resources, were to contribute as much as the lack of air support to the tragedy.

Freyberg, a lion in battle, was not the man to lead this complex defensive action; his analytical failings, understandable as they were, contributed mightily. This poor strategy and lack of preparedness would combine to frustrate the desperate and inspiring heroism of the British, Australians, New Zealanders, Greeks and Cretans who fought so hard and so well during the course of the battle.

CHAPTER 4

Fallen Flower Petals—Maleme, Chania and Rethymnon 20 May

The screaming Junkers over the grey-green trees,
Their cargoes feathering to earth.
They might be wisps of white rose petal
Caught in the keen, compelling twist of fate
Faltering, aimless, in an aimless wind:
Confetti, white and dirty white,
Tossed out in scattered handfuls ...
And one man idle, leans against the open hatch,
Through which the white horde poured
And watches Crete whine past below,
And in the mixed array of conquest
His hearing does not catch the rifle snap,
Sudden, faint his hands grasp deeply into nothingness,
And in bewildered agony
The dark soul drowns.

He struggles as the troop plane banks;
Unstruggling, falls in one slow turn -
The horror dream personified -
And the olives snatch him to their greenery.
Our vague ears do not catch the death – weak cry,
And someone blows the smoke shreds from his rifle mouth.[1]

It was shortly after 8.00 a.m. on the morning of 20 May. As ever in the eastern Mediterranean in the middle of spring, the weather was fine and clear with the promise of a very warm day.

Then from out to sea came a continuous, low roar. Above the horizon there appeared a long black line as of a flock of migrating birds. It was the first aerial invasion in history approaching. We looked spellbound.[2]

At first the Allied soldiers had thought the initial rumble of aero engines was nothing more than their daily dose of strafing arriving, but it quickly became apparent that this was something altogether more serious. As Private Peter Butler of 22nd Battalion, recalls:

> The morning of the 20th was fine and sunny and calm, there was no early morning strafing, we'd stood down, and while we were waiting for breakfast I was ordered to take the daily parade state to battalion headquarters on or near the top of the hill. I had on my web gear, no pack and carried a rifle. As I left the siren sounded but there was no sign of aircraft. About a quarter of an hour later the siren sounded again and I thought that was strange; there must be a big one coming. I couldn't remember being told this was the invasion alarm. By this time I was at head-quarters, and bombing and strafing started and really worked into quite a noisy show. This, I guess, went on for about twenty minutes and then stopped. There was a strange silence. I came out of the slit trench and looked around. Apart from a few fires on the 'drome there was no visible damage. Shortly, a distant roar of engines could be heard approaching from the north and then from over the sea came the sight of countless planes from as far east to as far west as could be seen, from horizon to horizon. The roar became louder and louder until they were overhead.[3]

Quite close, WO Les Young was at breakfast with the 21st:

> Breakfast was served at the usual time, I think 0730 hours and immediately after breakfast the men in my battalion were engaged in sharing out papers and parcels which had arrived late the previous night and which were addressed to personnel who had remained on the mainland of Greece. While this was going on a roar was heard in the sky and over came ME109s, Dorniers and Stukas and commenced strafing and bombing the area all around. There was a terrific din and the sky was black with planes. Apart from an odd bren gun I did not hear any ack ack [anti aircraft] fire. They were followed very shortly by the lumbering JU52s and ghostly gliders.[4]

Sweeping in at under 400 feet, beneath the elevation of the heavier AA guns, the Junkers kept in tight formation until they reached the drop zones, then the air blossomed with a blizzard of colour – pink or violet denoted an officer, O/R's black, weapons, white. This was

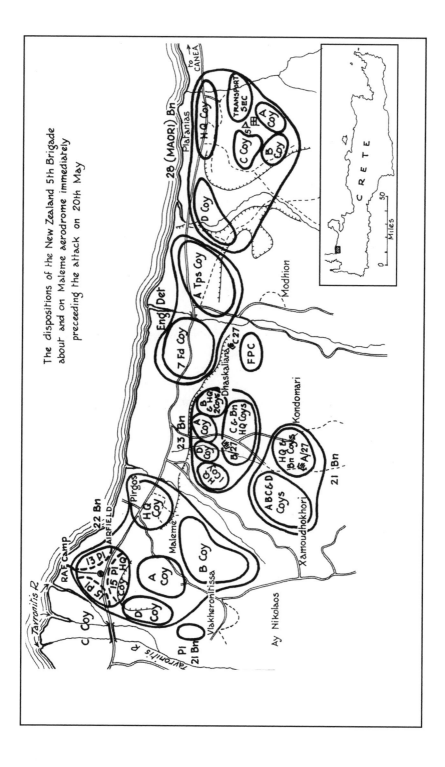

The dispositions of the New Zealand 5th Brigade about and on Maleme aerodrome immediately preceding the attack on 20th May a vast aerial armada, the like of which had never been seen before, a Wagnerian chorus of thundering engines and myriad planes, filling the vast blue of the eastern Mediterranean sky like giant, metal locusts, a plague; the judgement of God.

> [Freyburg] stood out on the hill with other members of my staff enthralled by the magnitude of the operation. While we were watching the bombers, we suddenly became aware of a greater throbbing in the moments of comparative quiet and looking out to sea with the glasses, I picked up hundreds of planes, tier upon tier, coming towards us. Here were the huge, slow moving troop carriers with the loads we were expecting. First we watched them circle clockwise over Maleme aerodrome and then, when they were only a few hundred feet above the ground, as if by magic, white specks mixed with other colours suddenly appeared beneath them as clouds of parachutes floated slowly down to earth.[5]

This vast airborne fleet, sweeping in from the sea to deluge Crete, was the culmination of Student's dream, the final working of his theories on vertical envelopment. The General in Athens might be heartened by the sight of his creation springing to awesome life but he could not be unaware of the risks; failure would, at best, end his career.

If the 3.7-in guns couldn't register, the Bofors, manned by the marines, certainly could and they fired till the barrels glowed red. The slow moving transports were a gunner's dream and the shells tore through metal and flesh, dismembering men and aircraft in mid air, slain parachutists tumbling 'like potato sacks' from the wrecked fuselages.

Hanging from telegraph poles, caught in trees, dumped in ditches, the dead bodies of *Fallschirmjäger*, the corpses swelling, putrid in the heat, a new feature of the island landscape. Was this to be the face of vertical envelopment? Scarcely what its creator had envisaged.

One of the prime weaknesses of the German tactical approach to parachute operations was the poor design of the harness. The lines were fastened to a webbing harness that held the wearer suspended as he dropped. He had the ability to use the personal weapons he had with him, but could not influence the course of his descent. Once the chute opened he was at the mercy of man and the elements till he touched down. From the ground, Charles Upham saw clearly just how vulnerable the *Fallschirmjäger* could be:

> The easiest sort of warfare in the world. If they landed where you
> were or within range of where you could get to they were just
> sitting ducks, they had no chance. Of course the whole object of
> parachute troops is to land them where there isn't anybody – out
> of sight. Once they get on the ground and regroup they've always
> got the very best of weapons and things like that. But while
> they're in the air – gliders and paratroopers in the air – oh,
> they're the easiest things in the world to bring down.[6]

The bombing and strafing that preceded the attack was over but
Stukas and Me109s still prowled the skies, ready to provide close air
support and pounce on targets of opportunity. Unknown to Student
a copy of the parachute training manual had been recovered from an
earlier assault the previous year and the defending troops were thus
versed in the principles of immediate counter-attack at every point of
landing to prevent concentration. This was the moment of maximum
danger for the parachutists, scattered, disorientated, separated from
officers and weapons.

If the defenders were at first stunned and awed by the power and
noise of the assault, they very soon recovered. Peter Butler continues:

> There was perhaps a minute's awestruck inactivity while people
> realised what was going on, then firing started from all over the
> area. Some paratroopers were firing machine pistols as they came
> down but this stopped quite a way from the ground and it
> appeared the majority were hit while still in the air. The gliders
> around us fared little better. They came in so low one could not
> miss. I saw one man firing a bren from his shoulder literally
> tearing a glider to pieces – bits were flying off it. It landed about
> twenty yards from me and only one man came out. He only made
> about two steps before he was cut down.[7]

This rapid response did not occur west of the Tavronitis where the
western flank of Major General Eugen Meindl's assault on Maleme
landed unopposed; Major E. Stenzler's 2nd, and Captain W.
Gericke's 4th Battalion of the Luftlande Regiment, accompanied by
two heavy weapons companies with anti-tank guns and mountain
howitzers; with them Lieutenant P. Muerbe and a commanded party
of a reinforced platoon whose job was to secure the extreme western

end of the island by taking Kastelli and the half finished airfield there.

Major O. Scherber with 3rd Battalion, divided into company sized groups and dropping along the line of the road between Pirgos and Platanias, comprised the eastern flank whilst the centre was left to glider-borne detachments under Majors Braun and Koch. The sedate, huge winged aircraft, swooping silently like prehistoric creatures, had spilled around the banks of the Tavronitis, ready to disgorge troops and their heavy equipment and guns. Their task was twofold; Braun was to silence the Bofors guns at the dry mouth of the river and secure the bridge; Koch to assault Hill 107.

As the lumbering transports began to appear over Maleme airfield, Major Rheinhard Wenning, in charge of the actual dispatch of the airborne troops:

> ... suddenly with a terrible sound my plane lurched forward and I could just see out of the corner of my eye that the right wing of the plane on my left was touching our wing. It could have been serious because obviously that pilot was concentrating on what was happening on the ground and didn't realise his plane had drifted into mine. We averted disaster by lowering our flap and banking to the right.

Then:

> As we brought the planes down to the required height of 150 metres above ground, and slowed to 160 km per hour we could see in front of us the first parachutes swelling out, then, in rapid order, more and more ... then Captain Wenndorf gave the signal for the jump by our unit by pointing a yellow flag through the roof of the cabin. Seconds later repeated shocks shook the plane, a sign that our parachutists were jumping out ... I looked down at the ground and saw one parachutist after another landing down there in an area which I can only describe as very difficult; mountain formations, very rugged and chalky, vineyards, olive groves, and dry fields, with small villages and narrow roads. Everywhere in this landscape were the countless white spots of parachutes, in trees or on the ground, where the soldiers had discarded them. And I could see parachutists moving into position and the first fighting erupting.[8]

Both Bofors and Brens should have taken a fearful toll, but it was mostly these lighter weapons that did the killing, shredding the fragile gliders. A deal of dispute hangs over the question of why the heavier guns at Maleme proved ineffective. There is consensus that the heavier AA guns did not fire at all. It may have been, as has been asserted, that the weapons were lacking parts, though other witnesses record test firing for several days prior to the invasion being successfully carried out.

A number of defects seem to have played a role: the guns were badly sited which made them relatively easy targets for the Me109s. The batteries were poorly camouflaged which added to their vulnerability and the arc of fire expected was too wide; better positioning of the individual pieces would have permitted a better concentration of fire.

These shortcomings were particularly acute at Maleme where the AA guns were deployed to cover the airfield itself:

> Since the effective range of the Bofors is little more than 800 yards, the gun positions were of necessity near the fringe of the aerodrome, and, in consequence most conspicuous and vulnerable. As soon as we were no longer able to operate from the aerodrome the role of these guns had changed. Their primary task should then have been to deal with troop carriers trying to land. They would have fulfilled their task much more effectively if they had been disposed irregularly at a distance from the aerodrome.[9]

Above the strip heavier guns were dug into the hillside. Their job was to act, not only in an anti-aircraft role, but also to cover the beaches – again the pre-occupation with a landing from the sea dominated defensive strategy. In the event these pieces proved useless as the barrels could not be depressed sufficiently to fire on the German transports, ideal targets as these were. The troop carriers came in, sticking to their low altitude, from the south, swinging in over the land to turn before the drop. No gunner can have had a finer target.

Any number of reasons, many of them sound, can be advanced as to why the available guns, which should have been more than adequate, were so badly sited: lack of transport, the difficulties moving the weapons under constant bombardment, but the primary reason was that the defensive planning was firmly fixated on an attack from the sea. To that extent Student's plan for vertical envelopment had worked. The concept was so breathtakingly audacious, it had clearly reflected its creator's genius for innovation.

Quite early in the confused and savage battle for Maleme, upon which the fate of the island garrison was to hinge, a spirited German attack from dead ground overran the Bofors, though the New Zealand infantry of C company held their ground and inflicted further, significant loss. D company's Brens riddled Braun's gliders, killing the major and many of his men. The survivors fell back toward the bridge where the lie of the ground afforded some cover. Here they were able to gain a lodgement and secure a precarious grip on the bridge itself. Some made it to the periphery of the airfield.

Koch's assault group fared no better. The gliders landed in two detachments straddling the opposing slopes but the fire of the defenders again accounted for many of them; those on the north eastern flank were also exposed to fire from 21st and 23rd Battalions. Koch too, became a casualty and was carried, dying, by the dazed survivors as they fell back behind the sheltering banks of the Tavronitis.

The 3rd Battalion transports swept in over the north coast in a roar of massed propellers, in the expectation the Bofors guns would have been taken. The aircraft were mercilessly raked, several shot down, many others damaged and the tight formation flying began to disperse. The parachutists, spewing in disorder over a four mile stretch of coast road, rock and scrub, landed directly on the positions held by 21st and 23rd Battalions, as Les Young laconically recorded:

> …the thought flashed through my mind how like the opening of the duck shooting season in New Zealand. The sky soon became full of parachutes in various stages of opening and descent and the shooting was good. I saw very little attempt on the part of any Hun to shoot back during his descent. Up to this stage I fail to see how any enemy could have landed unless wounded or killed. In fact it was a concentration of some of the best small arms fire I have ever seen. When targets failed to appear in the sky pot shots were taken at the doors in the sides of the Ju52s to try and get the first man before he made his jump. Good ack ack could not have failed to make a record bag. As it was, small arms fire seemed to be worrying both pilots and parachutists considerably as the planes would circle us, sheer off, and make a second attempt to land in the scheduled area.[10]

Scherber himself was amongst the many dead that littered the ground. Lieutenant Colonel Leckie, the Battalion commander, personally accounted for five of the intruders. A vicious firefight spluttered on for a couple of hours but the *Fallschirmjäger* were decimated and ruined as an effective detachment. A few managed to fortify houses below Modhion and hang on, but the attack was broken.

For the parachutists this was the stuff of nightmares; it would be immediately obvious to even the most dim witted that their intelligence estimates were grievously flawed and that the number and mettle of the defenders was infinitely greater than had been promised. As the men, already parched and sweating profusely in their wool kit (no allowance for tropical issue had been made), scrabbled for their weapon containers many were shot down or bayoneted before they fired a round. The survivors, leaderless, took to whatever cover they could find, in handfuls and rally groups, huddling in dried up stream beds, disorientated in the horticultural jungle of endless olive groves.

Sergeant Bill Ritchie found his morning routine unexpectedly interrupted:

[I] had just gone to the toilet and being in a paddock of grapevines in the open I was unable to get back to my unit and was certainly caught with my pants down. A parachutist dropped near me and all I had with me to deal with him was a spade; fortunately I got him before he got out of his 'chute. All hell seemed to be let loose as paratroopers appeared all around us. Corporal Brunning was seen to riddle a troop-carrying glider full of parachutists with his bren gun and the plan crashed at the back of Canea.[11]

German intelligence had completely misread the strategic realities which obtained on Crete. They seemed to have come to the wholly erroneous conclusion that the Allies were in the process of pulling out and that the campaign would therefore be a 'walkover'. Their awakening was to be most rude:

In view of the shipping movements in the first half of May, Air Fleet 4 considered it possible that British troops had been moved from Crete ... concerning the attitude of the population of Crete and members of the Greek and Cretan army, the High Command was of the opinion that there existed the possibility that influential circles desired the ending of the war and the extension to

Crete of the favourable terms which had been arranged on the mainland with the German forces. It was considered possible that the British, in view of the attitude of the population, were evacuating the island.[12]

Moreover, their intelligence had also misled them over the sympathies of the natives who rushed out, armed with whatever they could, and set about the invaders with a fury. Nazi Germany's Aryan elite were, in some cases, hacked to death by fearless Cretan housewives or blasted by ancient shotguns. To add to their dismay a re-supply flight of Junkers, cruising in during the late morning, dropped significant quantities of arms and ammunition, including machine guns, most of which were gratefully received by the defenders.

Eugen Meindl was forty-nine; old for a paratrooper. Like Student he was the living embodiment of the paratrooper ideal – fearless, resolute and always at the forefront. The position in which he found himself as the morning wore on into the hot, dry afternoon of 20 May was one that would have daunted any commander. His detachments had failed to secure any of their key objectives, they were scattered over the sector in penny packets, casualties were obviously dire, the strength, will and skill of the defenders formidable.

One of Andrew's most exposed forward units was D Company, less than seventy-five riflemen, bolstered by a couple of machine guns. Undeterred by the awesome parade of transports, the Kiwis shot down all those who were unlucky enough to land within range. Those landing west of the river were, however, relatively safe.

Stenzler's men, west of the Tavronitis had been able to form up relatively undisturbed and these represented the only intact formation on the field. Meindl therefore began to consolidate what he had, exploiting the meagre gains to dig in on the perimeter of the airfield. To try and retain some element of the initiative he ordered Stenzler to detach 5th and 7th Companies, swing around to the south and storm Hill 107 from this direction. It was a bold move and a desperate one, but both Meindl and his adjutant, von Seelen, were shortly afterward disabled by wounds. Gericke now took command.

The 22nd New Zealand Battalion had so far borne the main brunt of the onslaught and had given a first-class account of themselves. Nonetheless, as the morning wore on into the afternoon, their position became increasingly uncertain. Waves of Stukas screamed down upon their hilltop trenches, the harsh clatter of machine guns heralded yet another strafing run from a cruising Me109. The men were suffering from the stifling heat, exacerbated by a dwindling water supply.

One of Andrew's difficulties was the large and rather disorderly RAF camp. This lay at the extremity of the line held by C Company. It contained several hundred men, many of whom were unarmed. The difficulties in coordinating operations with the RAF meant that the colonel had little direct influence over these personnel and no proper defence of the camp appears to have been organised. Many of the ground crew were militant in their insistence they were not combat troops. Events have a habit of disregarding such nice distinctions, especially in the fog of war.

The paratroops, having secured the iron bridge over the Tavronitis, were now able to nibble around the exposed flank and penetrate the camp itself. No serious resistance was offered by the startled RAF who dispersed before these heavily armed figures, driven like game through the maze of tents. The collapse of any coherent defence enabled the Germans to work around both C and D Companies and to mount the beginnings of an assault on Hill 107. By noon, however, this potentially dangerous gap had been plugged and the advance halted. The laconic Lofty Fellows had been on an escort detail to the west of the strip:

> ... when the planes started to arrive so I decided to get down the bank in amongst some bamboo and I sat there and meditated about the comings and goings of aeroplanes and things – bombs went off at frequent intervals all around the place – dust and al that sort of thing – I thought, well, I wish they'd hurry up, because it was fairly early in the morning, and I can go and get some breakfast. But, they persisted in this damn bombing and unfortunately for me they dropped one fairly close and in the excitement I swallowed my cigarette and ended up topsy-turvy with a bit of shrapnel in my leg and the bamboo blasted apart, and I looked up saw all these bods dropping down in parachutes.
>
> Then I saw a glider crash land on a hill and I could see they were in force and I decided discretion would be the better part of valour and I should find some company, so I crept along through the bamboo and found a culvert and skipped through like a crab to the other side, and then dashing from cover to cover I arrived at a little machine-gun post underneath an old building. There was a Corporal Hosey. He didn't know what was going on, and I had no voice. I don't know if I was scared to death or swal-

lowing the cigarette had done it, and I was gesticulating and trying to point, and of course once he caught on and looked around he got a little dry in the throat too.[13]

Even more ominously the New Zealanders' forward positions were being shelled by the German guns and mortars safely landed west of the Tavronitis. Ammunition was also running low as were the batteries for their radios, communications became difficult and Colonel Andrew lost contact with his forward companies.

He could, however, still communicate with Hargest at Platanias. The troops huddled in their inadequate trenches on the summit and flanks of Hill 107 could not fail to notice that the weight of the German pressure was increasing and that the enemy deployment west of the river was proceeding without interruption.

The HQ Company of the 22nd was posted in and around the village of Pirgos. Lofty Fellows, by now back with the defenders and manning a machine gun, found a unit of paratroops forming up, directly under his sights, apparently oblivious. He very soon made his presence felt. Still in need of a well-earned breakfast he also managed to acquire a 2-inch mortar which he utilised to suppress a further body of the enemy, bunkered in one of the houses. The deadly game of hide and seek continued through the morning till virtually all of the lodgements had been mopped up, save for the inevitable crop of sharpshooters.

Not entirely without hindrance, one detachment west of the iron bridge was in serious difficulties, Lieutenant Paul Muerbe and his seventy-four paratroopers who had been tasked to mop up the outpost at Kastelli. This was manned by the 1 Greek Regiment, nominally 1,000 strong but with barely sufficient arms and those of indeterminate quality for every other man and less than a handful of live rounds apiece.

These local recruits were stiffened by detachments of the gendarmerie, earlier animosities submerged, and led by Major T.G. Bedding with a small cadre of New Zealanders as instructors. The Major had divided his command into two weak battalions posted either side of the town. Muerbe's company were few but they were heavily armed and bolstered by the full, Aryan bravado of the *Fallschirmjäger*, they were scheduled for re-supply late morning and were not anticipating serious resistance. In this they were to be disappointed.

They lacked arms, ammunition and training but no one could say the men of Crete lacking fighting spirit, it coursed through their veins, they had not run from invaders in the past and they were not proposing simply to hand over their town, their homes and their honour. Muerbe's men dropped in two sticks, both to the east of Kastelli and they had jumped onto the bayonets of Bedding's first or 'A' Battalion.

It is said many Germans were shot as they floated to earth,[14] others riddled as they attempted to fumble clear of their harness, still more stalked and sniped by the locals who knew every blade of grass and the lie of each dry bed and ditch. Each German killed yielded a cache of superior weapons. Bedding and his command group put their Brens to good effect.

Disorientated, the surviving Germans began shooting wildly, often into their own men. As they stumbled, dazed and leaderless, through the maze of olive groves, Cretan fighters, men and women, would rise like shrieking spectres from each fold of cover and set about them. This was warfare at its most crude and bloody, no quarter given, clubs and knives hacking down the pride of the Luftwaffe.

By mid morning the remnants of Muerbe's detachment were frantically fortifying a cluster of farm buildings as a last ditch defence. Having contained and emasculated the attack Bedding now wanted simply to surround and contain. This was too tame for the Cretans who launched a wild and desperate charge, pressing the attack home despite heavy loss and dealing with Muerbe's survivors. Few came out alive, those who did were incarcerated by Bedding in the town gaol, primarily to save them from the bloodlust.

The subsequent German report on the action concluded that:

> The platoon commanded by Lieutenant Muerbe which was put down east of Kastelli immediately became involved with strong guerrilla bands and Greek troops, strength about a battalion, under British commanders, and was mopped up in the course of the fighting. Of seventy-three men, twenty wounded were liberated later. Lieutenant Muerbe and fifty-two men were killed. The majority were found to have been grievously mutilated.[15]

The Cretans, with their long history of uncompromising resistance to the invader, were quick to adapt one of their traditional warrior songs:

> Where is February's starry sky
> That I may take my gun, my beautiful mistress,

> And go down to Maleme's airfield
> To capture and kill the Germans.[16]

Whilst some of the German dead may have been cut up by the locals, it would appear more likely that damage to the bodies was simply the work of predators. Such assertions, however, would serve to fuel resentment against the indigenous population and be employed as a rationale for the savagery of reprisal.

Colonel Leckie, commander of 23rd Battalion, found himself sharing the experience of having Germans landing, in his case, almost literally, on top of him. The Colonel immediately brought his Thompson into use, to good effect. His adjutant, otherwise immersed in regimental business, accounted for a pair of the intruders without being constrained to rise from his improvised desk. His experiences were similar to those of the 21st and 28th who accounted for most of the intruders in their respective sectors. The consequences for the parachutists were grim, Major W.D. Philip, a gunner in charge of an artillery unit recorded:

> The troop riflemen are still below ground and so we raise them and organise them along the front edge of our position. After the first excitement one Hun is only about 25 yards away among grape vines. A few rounds are fired but he may be lying 'doggo'. Gnr McDonald sets our anxiety to rest by coming up from the opposite direction, walking straight up to him and saying, 'You'd look at me like that, you bastard, would you?' with appropriate action. Another poor devil gets his on the wing. His 'chute catches in an olive tree and he finishes up by leaning on a rock wall head on hands as if he had been meditating by the wall when death caught up with him. Dead Germans everywhere.[17]

In the centre the attack toward Chania was in difficulties before Crete was ever sighted. General Sussman, together with his HQ staff, was killed when his glider folded, not long after take-off, and smashed into the rock strewn hillsides of Aegina. The baton of command thus passed to Major Richard Heidrich, commanding 3 Parachute Regiment. The plan for a pincer movement from dropping zones on the flanks was to be supported by glider-borne attacks on the defenders' AA batteries.

An Allied officer in Chania later recalled:

> All over the town everyone was letting loose at the planes,

throwing at them everything we had got. In the house next to us a section responsible for the repair of light ack acks, assembled a Lewis that was in for treatment and were firing it hard from their attic window. Out in the open on a terrace below the house was a major with his tin hat on the back of his head using a rifle for which he'd raised some rounds of tracer ammunition. You can see that way, he pointed out, by how much you are missing the things. Their technique was quite interesting. They put over a big bomber flying very low, dropping a stick which made us keep down and another plane followed behind us in a matter of seconds and dropped the parachutists where the bombs had dropped.[18]

A sapper from the New Zealand Field Park Engineers looked on as the circling gliders found out just how tough their reception would prove:

... several gliders heading in our direction and one in particular was making for the town below me and I began to wonder where he intended to land as the only open area was where the underground shelters had been made. In a wonderful piece of flying the pilot dropped the glider down right on it – it was a drastic finish for his passengers who did not last long, but they sure had guts to try a stunt like that ... one glider was about to land on top of us when the pilot must have seen the posts carrying the grapevines and flew on another two hundred yards, where he crash landed on a rocky area at the head of our gully. The glider went up in smoke and the crew suffered the same fate as the other poor devils.[19]

Although the defenders, well versed in the techniques of opposing an airborne assault, reacted with ruthless vigour it was difficult not to admire the bravery of the *Fallschirmjäger* who pressed on to certain death with such élan. The men who had earned such a formidable reputation were undergoing their greatest test. In the main they would not be found wanting.

Captain Altman's northern detachment, fifteen gliders of 3 Parachute Regiment, were tasked to silence the guns on the Akrotiri Peninsula; despite losses some of the group attained their target only to find the guns, reputedly sited near the Venizelos tombs, were dummies.

The Northumberland Hussars, the 'Noodles', were dug in around

Profitilias on the Akrotiri Peninsula, (or as dug in as the surface rock would allow). Sir Lawrence Pumphrey had landed at Souda after the regiment was evacuated from Greece. He was part of C and D squadrons (A & B had been taken directly to North Africa).

Since arriving on the island the Hussars had enjoyed the warm Mediterranean spring. Sir Lawrence, with his brother John, also serving in the regiment, spent many hours walking over the broad expanse of the Peninsula, visiting the small villages and the ancient monasteries that seemed to grow from the living rock. Nonetheless their positions were so well constructed as to be almost invisible from the air and German aerial reconnaissance had failed to detect their presence. Both sides received a substantial shock on the morning of 20 May:

> The [Germans] were dressed in grey boiler suits, rubber soled boots and brimless steel helmets, and were wonderfully equipped. They were armed with Tommy guns, chiefly, though some carried telescopic-sighted snipers' rifles and machine pistols. Each glider had a large quantity of hand and stick grenades.... On this day the armament of the Northumberland Hussars on the Isthmus comprised either a .303 English or .300 American rifle to each man, but no bayonets; fourteen Bren guns; and .38 revolvers for the six officers ... .[20]

There was no armoured or artillery support and the battalion possessed only one grenade, though this was put to good use. The Hussars, originally a Yeomanry regiment, had been deployed in Greece in an anti-tank role, where the Duke of Northumberland served as a troop commander. Despite fighting a series of successful rearguard actions, most of their heavier equipment had had to be abandoned in the evacuation.

Six German gliders came down, virtually in a line, in front of the olive groves where the HQ company occupied prepared positions; a further quartet of the intruders landed on or around the dummy AA position. (Although the deception succeeded, the sandbagged emplacements left behind enabled the paratroops to establish themselves and, for a while at least, hold out.)

Of the half dozen, virtually all were accounted for on landing; a number of parachutists were captured, many more killed. Those who did land and get out were swiftly counter-attacked. One glider, the second, blew up when rounds detonated the grenade locker. Only the crew of the third or 'C' glider managed to get free and dig in. The fight to destroy their foothold was bloody, costing the Noodles a number of casualties. Major David Barnett, the second in command, leading a probe in the thicket of olive groves, creeping up to a low wall, carefully raised his helmet on a stick to draw fire. None came, but the instant he lifted his head he was drilled by a sniper.

Private Arnold Ashworth was one of those who took part in this fight; he used his new Lee-Enfield, still in its protective daubing of grease, to good effect. Advancing over the dense ground he heard a wounded German crying out for someone to end his agony. He found the casualty in 'a little grassy hollow, a few yards from his comrades. I don't know he had got so far, for half his hip was shot away. He had been hit with the heavy calibre bullet of an anti-tank rifle … A short while before he had been a fine specimen of manhood … and now here he lay at my feet pleading with me to put and end to his horrible suffering and wasted life.'21

In such circumstances it was impossible not to feel pity for such a desperately wounded foe, the dead enemy lying broken and bloodied amongst the pathetic remains of their personal effects and family photographs. Such a sight tends to evaporate the demonic quality of even the most fanatical opponent and Ashworth, like so many others in similar positions, was bound to wonder if someone soon might be viewing his remains in a similar situation.22

The survivors of 'F' glider also managed to gain a brief lodgement but the single cherished hand grenade was thrown in a dynamic counter-attack which resulted in the death or capture of all the parachutists. Altman's attack, had it achieved success, could have threatened Creforce HQ, destroying command and communications structures.

The AA guns sited a mile or so south of Chania, at a crossroads on the Chania Mournies Road, were to be attacked by Lieutenant Alfred Genz and his company landing in nine gliders. Genz was already a worried man. Just as his glider began to move on take-off he received a belated intelligence message that intimated the garrison he and his comrades were about to attack was at least three times greater in numbers than had been anticipated. This was scarcely an auspicious beginning. As the descent began his worst fears were realised when two of his gliders crashed over Chania and another exploded when anti-aircraft fire detonated a box of hand grenades.

The remainder were able to land close to the battery manned by a single troop from the 234th Heavy AA Battery operating four 3.7-in guns. Despite a spirited resistance Genz and his men successfully overran the position; few of the gunners survived. Glider-borne troops possessed the inestimable advantage of having all their weapons, including MG 34s, with them and could thus swing fully into action immediately on landing.

He now proceeded to his second objective – the nearby wireless station – but was repulsed by other elements of Force Reserve, 1st Ranger Battalion and more Bren Carriers of the 1 Royal Welch Fusiliers. Isolated and without support Genz clung briefly to his gains but wisely opted to break off and attempt to break out toward the rest of his regiment. Luck, cunning and their commander's excellent English got them through.

The bulk of Sussman's troops were to be landed in the dead ground of Prison Valley whilst Heilman's 3rd Battalion 3 Paratroop Regiment was to drop at Karatsos and secure the coast road. The two forces would then converge on the capital, the main body on the Alikianou – Chania axis and the lesser along the axis from Platanias. The plan began to unravel when the transports, sailing majestically over the Akrotiri Peninsula, were filleted by heavy AA fire; the surviving parachutists scattered over a wide area, several battalions became hopelessly intermingled:

> The moment we left the planes we were met with extremely heavy small arms fire. From my aircraft we suffered particularly heavy casualties and only three men reached the ground unhurt. Those who had jumped first, nearer to Galatas, were practically all killed, either in the air or soon after landing ... approximately 350 men of my battalion survived the initial landing and organising period.[23]

These tough, young German *Fallschirmjäger* proved, in some cases, to be far from their heroic best in these chaotic conditions. Whilst many fought stubbornly in small groups and refused all quarter, others gave up without a fight and in the bloody scrum of the mêlée it was the New Zealanders and their Greek comrades who fared best. The paratroops were trained to form up swiftly on landing, follow their officers and fight hard to secure an identifiable objective.

As was often the case on Crete, officers and NCOs were missing or dead, the drop zones so scattered that it was nigh on impossible to determine where the actual objective lay and the quality of the opposition exceeded all expectation. The men who dropped on Eben

Emael were anticipating a hard and desperate fight, their objective was impossible to miss, whereas here they'd been promised a walkover. The reception they received must have come as a very unpleasant shock:

> ... the noise of the screaming plane engines, the sirens on the Stukas, exploding bombs, cannon and machine gun fire .. as the straffing continued large gliders uncoupled from their towing planes and settled silently on the far side of the Prison Valley, quickly disgorging their troops and heavy weapons. The gliders were followed immediately by the arrival of groups of big Ju52 transports, spilling out masses of white and multicoloured parachutes. Cracking of rifle and Bren gun fire broke out both from the invaders in the air and from our troops among the olive groves.[24]

Heidrich's three battalions, together with his engineer battalion and some heavy weapons companies, had landed without too much difficulty in Prison Valley but again their intelligence had let them down for, rather than finding themselves commanding a plateau, they were dropped into the bowl of a shallow valley with the heights before Galatos held by the defenders.

The 4 and 10 New Zealand Brigades had given Heilman's battalion a very rough reception. Those who dropped around the positions held by 18th and 19th New Zealand Battalions with the 6 Greek Regiment were shot down in scores; those who came down further south did succeed in linking up with their comrades in von der Heydte's 1st Battalion. At least one section of parachutists dropped into the water of the Alikianos reservoir, wherein all drowned.

Corporal Howard of the 18th used tracer to direct his fire at descending paratroops, the air alive with the crackle of small arms. Parties of Germans were coming down barely 200 yards distant where they provided excellent targets, many being dropped as they struggled to free themselves from 'chutes that swiftly became shrouds. In the same sector Lieutenant Colonel Gray:

> ... saw a parachute hanging in a tree and detected a movement round the left side of it. Fired quickly with my rifle then advancing very softly and quickly up to the parachute I looked around the side to see a Hun lying on the ground beside a gaily coloured container. He moved so I shot him at once to make sure,

and then moved cautiously from cover to cover ... I saw George Andrews sitting on the ground taking careful aim at some cactus bushes behind us. 'Steady on George,' I said, 'You will be shooting one of our own chaps.' 'No bloody fear, it's a Hun!' he said, and fired. 'Got him.'[25]

Upham's company found themselves pinned down by the supporting strafing but were able to launch minor raids against the concentration of parachute and glider-borne invaders, the men returning loaded with spoils in true brigand fashion. The enterprising Major Thomason put his ill-gotten gains to sound practical use:

About four hundred parachutists were landed right on top of us and were suitably dealt with. I gained much information from a German officer who I wounded in the foot with my revolver on the way down ... I found in his pack a fairly large Nazi emblem and asked him what is was for and he said: 'If we require supplies we expose this'.[26]

Putting this intelligence into effect Thomason received a handy drop of equipment, including rations and weapons.

The situation generally, where the invaders landed in defended areas, became very confused, with sporadic firing erupting like wildfire; a deadly game of hide and seek in the tangled terrain. Friend might fire on friend in the confusion but, as the morning wore on, the surviving pockets of attackers were gradually mopped up.

Landing without incident, the laconic Heydte had experienced a relatively easy landing with the distinctive white walls of the prison clearly visible in the distance. His quiet enjoyment was abruptly shattered when an Me109 took a particular dislike to him and he was obliged to leap into a ditch to avoid the hail of bullets, the irony of being first shot at by his own side did not escape him.

...[the fighter pilot] obviously never imagined that this lack-adaisical figure wandering about in such unmilitary fashion down the centre of the road could possibly have been the commanding officer of a German battalion.[27]

Parachute Engineers coming down by Episkopi fell foul of the Greek 8 Regiment who offered them a very warm reception. When von der Heydte combined with Derpa's 2nd Battalion, the paratroops overran and took the village of Aghia where Heidrich established his

HQ, joined by the diminished remnants of the divisional command whose four gliders landed foul of hidden tree stumps not revealed by aerial reconnaissance, suffering yet more casualties. Von der Heydte, as he set up his HQ, and from which the distinctive profile of the church at Galatos could be seen, noted by mid morning:

> When the first reports arrived from the companies they were contradictory and obscure. While sections of the battalion, advancing westwards on the right flank, along the high ground south of the valley, had met without any serious resistance from the enemy, the resistance in the valley itself had increased by the hour. The British seemed slowly to be organising a line of defence. During the first hours the fighting had apparently consisted of single, disjointed actions, sudden unexpected encounters, and mutual surprises; but now both sides were gradually organising themselves for battle.[28]

Puttick's failure to heed Kippenbburger's warnings of the tactical significance of Prison Valley had just been exposed. The respite gave the Germans an invaluable breather, to halt and reorganise their badly mauled forces. This fog of confusion and loss that attended the dispersed landings had robbed Heidrich of an immediate opportunity to mount any serious probes against the defenders' positions around Galatos. Local attacks were pressed home with vigour but, lacking coordination and support, were largely doomed from the outset:

> We advanced to attack [Cemetery Hill]. We proceeded, without opposition about halfway up the hill. Suddenly we ran into heavy and very accurate rifle and machine-gun fire, the enemy had held their fire with great discipline and allowed us to approach well within effective range before opening up. Our casualties were extremely heavy and we were forced to retire leaving many dead behind us ... the first attack on Galatos had cost us approximately 50 per cent casualties about half of whom were killed.[29]

These defenders were not, in fact, crack infantry but a scratch formation, the Petrol Company of the Composite Battalion. Desperately short of rifles and machine guns, the men were mainly support personnel:

> ... the rifles were without bayonets, and five fewer than the men

who needed them, and besides rifles there were only two Bren guns, one Lewis machine gun and an anti-tank rifle. The men were for the most part drivers and technicians and so ill trained for infantry fighting.[30]

Despite the losses inflicted and the failure, on the Germans' part, to attain any major objectives, they had been allowed to concentrate in Prison Valley. This was a serious omission on the Allies' part. That the area was suitable as a drop zone had not gone unnoticed. As Captain Lomas observed:

> ... noticed the unusual attention the German planes paid to the wide Prison Valley, a level area with plenty of good cover, dominated by the fortress-like prison which would be an impregnable stronghold for airborne enemy forces. Major Sean McDonagh, OC Petrol Company, was also disturbed by the fact that the chief Warden of the prison was fluent in several languages, including German, and we were convinced that landings in force would occur in the valley with the prison as an essential strongpoint.[31]

Despite the obvious potential, Puttick appeared oddly reluctant to provide for a more concerted defence of the valley. Kippenberger was, on taking command of 10 Brigade, immediately struck by the scale of the risk:

> Colonel Kippenberger took command of the 10th Brigade on 14th May and was soon aware of the shortcomings of the Greek regiments and the lack of infantry training of the ASC and Divisional Cavalry troops. He remonstrated in vain with General Puttick, going so far as to rub off the circles indicating the Greek positions on the General's map, insisting that they could not be considered effective military formations, but for all his pains he was met with 'Don't spoil my map Kip.' The 20th Battalion, in reserve, could well have supported or replaced the Greeks.[32]

A savage incident occurred when the 10th Parachute Company, under Lieutenant Nagle, dropped onto the reserve areas occupied by medical units, 7th General Hospital and 6th Field Ambulance. Lieutenant Colonel Plimmer at once surrendered but was killed along with twenty or so patients, presumably those incapable of walking; the rest were apparently dragged to their feet and herded in front of the attackers as a human shield.

They came up against 18th New Zealand Battalion at Evthymi where their attempts to break through toward Prison Valley generally resulted in death or capture. A subsequent enquiry dismissed the concept of an atrocity, blaming the incident as caused by the 'fog of war' – it does appear the hospital patients were marched as a POW column rather than a screen.

Prior to the attack, as noted, Brigadier Kippenberger had protested to Puttick that the disposition of forces covering Prison Valley was inadequate. He was most concerned that the Greek regiments and the ancillary New Zealand forces were neither fully trained nor equipped. He went so far as to assert that the Greeks were so unprepared, their deployment was tantamount to murder. In fact his concern over the mettle of the Greeks was unfounded but his doubts overall were sound. The Germans, though mauled, were still able to concentrate their surviving forces without hindrance.[33] This was to prove significant.

A particularly glaring omission, for which some writers have blamed Kippenberger, was the failure to maintain an adequate garrison in the prison itself. The building was a veritable fortress and the fact the Germans were able to move in unhindered, gave them a strong tactical HQ.

> In the third sector attacked, namely Rethimnon, parachutists at first captured the aerodrome, but were wiped out by a spirited counter-attack by Greek and Australian troops belonging to the 19th Australian Infantry Brigade.[34]

This description, from the Official History, is something of a simplification. The situation at Rethymnon was both more complex and more protracted than is suggested here. The Allied forces, in this sector, were commanded by Lieutenant Colonel Ian Campbell, a relatively young and energetic officer, who had only recently been appointed to command.

Campbell entrusted the defence of the town, with its mighty Venetian fortress guarding the harbour, to the 800 odd Cretan police, under Major Christos Tsiphakis, later to become a key figure in the resistance. The town was, by now, out of bounds to the Australians whose hard-drinking, free-fisted ways had occasioned a number of disturbances.

Major Alfred Sturm, leading the airborne assault, commanded a powerful force with which to achieve his objectives. These were, broadly, to secure both the town and the airstrip and then march westwards to consolidate the main group centre target of Chania/Souda. For the assault his second Parachute Regiment of three full and well-supported battalions was divided into three commands.

Captain Weidermann led the 3rd Battalion with two detachments of airborne gunners armed with howitzers, some mortars and anti-tank weapons, supported by a machine-gun company. These were dropped over the village of Perivolia and tasked with storming Rethymnon. Sturm, with a beefed up assault company and his HQ, was to land around the airstrip, whilst Major Kroh with 1st Battalion came down to the east and their job was to march west and ensure the airfield was secured. Kroh had additional heavy weapons including flame throwers, motorcycles with sidecars and more machine guns; a most potent combination.

Against this, Colonel Campbell could only count on two Australian battalions; the 2/11th (West Australians) under Major Sandover and the 2/1st, Campbell's own unit. The heavy weapons at their disposal were vastly inferior, both in quantity and quality, to those of the attackers. Their artillery comprised four antiquated Italian 100-mm guns and the same number of US 75s (minus sights).

If the defenders lacked numbers and adequate matériel Campbell, in part, made up for this by concentrating his defence around the airstrip and leaving only a light screen around the town and on the beaches. Topography also lent a hand as the airfield was dominated by two hillocks, A on the east and B on the west, which straddled the Wadi Pigi and dominated the level ground below. Campbell placed his two battalions accordingly, 2/1st on Hill A and 2/11th on Hill B. Between the two eminences he distributed the 4th Greeks, with the 5th further back in the further range of shallow hills around Adhele village.

Campbell had concealed his two Matilda Tanks in a dry river bed west of the airstrip and set up his own HQ on Hill D. This enabled him to remain in touch with the rest of the defenders and to respond quickly to any crises which might arise.

The very lack of heavy weapons and anti-aircraft guns, combined with the skilful camouflage of the defenders' trenches, succeeded in fooling the Germans into believing the area was very lightly held indeed. Captured aerial photographs showed the enemy had located only one of the defensive lines and this was immediately altered.

Determined to give the enemy no inkling of his strength or disposition, Campbell had only permitted the men to break their cover when they went swimming in the warming waters. No more than a score of men at a time were allowed this luxury. The returning swimmers then had to earn their privilege by attempting to creep back unseen, thus alerting Campbell and Sandover to all possible routes an attacker might take and where the areas of dead ground lay.

As Sturm's drop formed part of the secondary afternoon attack on 20 May there was no question of surprise and the inevitable ground strafing and dive bombing was less thorough and sustained than elsewhere. After a quarter of an hour the unmistakable sound of the heavier transport engines could be heard and the first two dozen Ju52s appeared, flying eastward in perfect formation along the coast. More appeared piecemeal and the lack of cohesion amongst the badly stretched squadrons was obvious.

Despite the fact that the drop was part of the second afternoon wave, the defenders had not been made aware of the morning landings further west. In consequence a number of officers were absent from their posts. Nonetheless, a terrific hail of small-arms fire greeted the intruders, the Brens hammering wildly as fast as the magazines could be loaded and these accounted for at least seven of the 160 odd planes that appeared, whilst another two exploded in a mid-air collision.

Some of Kroh's men fell into the Mediterranean and were swiftly pulled under by the weight of their equipment. One of the Australian defenders recalled that:

> I was spellbound by the futuristic nature and the magnificence of the scene ... they were coming in along about five miles of coastline and as far as the eye could see they were still coming. They were about 100 feet above the water and rose to about 250 feet as they came over the coastline, dropped their parachutists, dived again and turned back to sea.[35]

Of the 161 transport planes involved in the drop, fifteen were brought down by the Allied guns, one of whose gunners:

> ...saw planes burst into flames, then the crews inside feverishly leaping out like plums spilled from a burst bag. Some were burning as they dropped to earth. I saw one aircraft flying out to sea with six men trailing from it in the cord of their 'chutes. The

'chutes had become entangled with the fuselage. The pilot was
bucketing the plane about in an effort to dislodge them.[36]

On one of the planes, the platoon commander, first to jump, was
killed before he exited. His comrades were so unnerved they refused
to budge. The pilot swept around for a second run but this time the
plane was riddled with one of the engines catching fire. He ditched
in the sea and the shaken survivors came ashore by rubber dinghy.
The continuous hail of fire accounted for all but a pair of them.
Another group who jumped as instructed died horribly when they
fell into a cane-brake, becoming impaled on the lethal bamboos as
surely as on the bayonets of the defenders.

Those who were dropped near the airstrip, literally under the guns
of Sandover's defenders, were massacred. Many landed as lifeless,
bloodied bundles. Sturm himself survived, being lucky enough to
come down in some dead ground, but most of his men were killed or
scattered, the defenders' Vickers machine guns firing till the barrels
overheated and jammed. Such was the confusion in the air that
sticks of paratroopers were being dropped at intervals over ground
already carpeted with dead, particularly on the eastern flank of
2/1st's positions astride Hill A. That night Sandover led his men
forward to clear the field. Eighty-eight shocked and wounded
captives were rounded up, together with quantities of equipment.
Sturm was taken the next day.

Despite their appalling losses the sheer weight of numbers dropped
around the airstrip enabled the Germans to maintain their attack on
Hill A which, after stiff fighting and with the Vickers all out of
action, was taken. As much of the heavier arsenal had been dropped
here, the paratroops were well equipped with mortars and anti-tank
rifles. A counter-attack was put in at dusk on that first day with the
tanks going in on the left flank but both became bogged down. The
infantry was held up by determined fire from the attackers who had
now infiltrated the western slopes and established positions in the
olive groves on that flank of the connecting spur.

This was the only serious foothold. Kroh's companies further east
were dispersed and depleted, being driven back almost to the beach,
digging in as best they could in the dunes. Those who'd managed to
establish positions at the base of Hill B were dislodged at dusk by
counter-attacks and driven into the network of vineyards around
Perivolia. Many were captured and, from a dead officer, Campbell's
men salvaged a copy of the Germans' signal code. This was put to

good effect the next morning when, once Major Sandover had translated, the defenders summoned up a re-supply of small arms and ammunition. With typical efficiency the Luftwaffe obliged.

At the end of the confused and bloody afternoon's fighting the only German gain was the seizure of Hill A, which Campbell was determined to re-take. Kroh, having gathered such additional survivors as he could during the balmy Mediterranean night, was equally bent upon pushing out across the spur, his patrols having driven in some of the Australian picquets.

In the thinning light of dawn, as the defenders massed and advanced, they ran into the attackers coming the other way. With their telling superiority in automatic weapons the Germans quickly gained fire supremacy and inflicted casualties on the Australians. As the firefight developed groups of parachutists began to nibble around the flanks of Campbell's attackers. Captain Moriarty who commanded the reserve of infantry and supporting Bren carriers, reported that the situation had become critical.

Realising the crisis of the battle was at hand Campbell threw in his final reserves who were able to approach unseen along the dry bed of the Bardia stream. Moriarty was ordered to resume the attack with his swollen force which he divided into four and, avoiding the main German concentration dug in on the summit of Hill A, began to filter around from the north, making full use of the cover afforded by the maze of gullies. Apparently under fire from all sides and duped into thinking Campbell's forces were considerably more numerous than was in fact the case, Kroh's resolution began to waver. Being murderously strafed by his own air support that killed nearly a score of men, clearly did not help faltering morale.

By noon the fight was effectively over; a significant number of paratroops surrendered with the loss of much additional matériel. Some were taken while attempting to disguise themselves as Greeks and the haul included Major Sturm himself, captured without ever having been able to influence the course of the battle.

Kroh was not yet done. He and those who had avoided the Australians' round up, retreated to a substantially built olive oil plant at Stavromenos to the east which was already occupied by the heavy weapons contingent. With determined effort and the genius of the German soldier for constructing detailed scratch defences, the factory was turned into a fortress. This was not just a last ditch action but a means of containing the victorious defenders who, without heavy artillery, were at a severe disadvantage in what would amount to a siege.

On the western flank the surviving paratroops had made no effort to influence the fight for Hill A, preferring to consolidate their own precarious position by digging in amongst the houses of Perivolia and turning St George's church into a second fortress. On the 22nd the 2/11th drove into the German outposts and began clearing the houses. Using their captured signals they galled the defenders by arranging for Stukas to dive bomb the village but the bastion of St George's Church, which commanded an all round field of fire and provided, from the tower, excellent observation, proved a very tough nut indeed.

The attackers' toehold in Perivolia had the same effect to the west as their tenure of the olive plant provided in the east. Though the airborne assault had completely failed to attain any of its specified targets, the hard pressed and much depleted survivors were effectively able to contain the garrison on both flanks.

Campbell remained determined to see the intruders off altogether and, on the morning of the 22nd, he'd ordered an attack on the redoubt at Stavromenos. The 75-mm guns however, proved utterly unable to penetrate the thick stone walls and the valiant Moriarty was shot dead by a sniper. An infantry assault was planned for the evening when it was intended that the Australian attack from the west would be supported by 200 Greeks coming up from the south. In the event the latter did not appear and the defenders were able to hold their ground. With their superior firepower and Kroh's incisive leadership the defenders were strongly placed; the fight was far from over.

As the main battle for the island raged in the west, Campbell, despite his limited resources, kept a relentless pressure on the German positions. A brace of captured anti-tank guns was turned against the stout walls of St George's, armour-piercing rounds smashing into the crumbling masonry. By the 27th, after Freyberg had decided to throw in the towel, the ruins were taken by assault and the tanks brought up to support the final assault on the remaining German positions in Perivolia. The ageing Matildas proved unequal to the task, proving chronically unreliable, but the fight was continued with vigour:

[One platoon had broken into the German position but was now cut off] ... that left me only one thing to do – attack to help Roberts out of trouble or to complete the success he had started.

I knew I'd have to lose men, but I couldn't lose time. A section from 14 platoon, nine men, was ordered to move to a low stone wall fifty yards ahead around a well about twenty-five yards from the German front line, to cover with Bren fire our attack across the open. They raced along the low hedge to the well. The leader, Corporal Tom Willoughby, was nearly there before he fell. The man carrying the Bren went down. Someone following picked it up and went on until he was killed and so the gun was relayed until it almost reached the well in the hands of the last man, and he too was killed as he went down with it. Eight brave men died there – Corporal Willoughby, Lance Corporal Dowsett, Privates Brown, Elvy, Fraser, Green, McDermid and White. The ninth man, Private Proud, was hit on the tin hat as he jumped up and fell back stunned into the ditch.[37]

The day before, the factory at Stavromenos had finally been taken after the 75s were brought up to fire at point blank range. The sheer grit and determination shown by these Australians, well and decisively led, ensured that Rethymnon did not become a second Maleme. That the island was eventually ceded in defeat does not detract from their success. The example of Rethymnon also shows how a vigorous and well coordinated defence, with immediate local counter-attacks being pushed home with skill and determination, could achieve against paratroops.

In a detailed study of the fight I.M.G. Stewart observes:

Unprejudiced by memories of the First World War he [Campbell] had launched his men upon immediate counter-attack. From every side his troops had leapt upon those Germans who had reached the ground alive inside the lines of defence. Within an hour he had committed his reserve, sending his tanks into the open across the airfield. Here was a calculated risk bravely taken … there was no time for half measures. Already the issue was for all or none.[38]

There is no escaping the conclusion that an injection of the same urgency and dash could have turned the tide at Maleme and thus altered the entire course of the battle. That, perhaps more than anything else, represents the tragedy of Crete.

CHAPTER 5

Stones and Sheath Knives—
the Crisis 20/21 May

I was spell-bound by the futuristic nature and the magnificence
of the scene before me. It wasn't long before they were coming
in along about five miles of coastline and as far as the eye could
see they were still coming. They were about 100 feet above the
water and rose to about 250 feet as they came over, dropping
their parachutists, dived again and turned back to sea. I saw
many Huns drop like stones when their parachutes failed to
open. I saw one carried out to sea trailing behind the plane with
his parachute caught in the tail. The men all had black 'chutes;
ammunition and guns were dropped in white ones.[1]

Although Chania remained the administrative centre, Heraklion,
roughly half way along the north coast, was the largest town with
the best docking facilities. Like the other coastal townships it was
Venetian in character and, just inland, lay the Bronze Age Palace
complex of Knossos, made famous by Sir Arthur Evans, whose
partial reconstruction continues to divide academic opinion.

Like Campbell at Rethymnon, Brigadier Chappel at Heraklion had
taken the wise decision to concentrate the bulk of his available forces
around the airstrip, the largest of the three. In terms of men he was
better placed than Campbell but shared the same deficiencies in
heavy guns and mortars.

As well as two Australian battalions, the 2/4th and 7th Medium
Regiment, (deployed as footsoldiers) and three Greek formations he
commanded three battalions of British regulars – 2nd Black Watch,
2nd Yorks and Lancs, 2nd Leicesters. Like Rethymnon the airfield
and port beyond were covered by a pair of shallow hills just over half
a mile inland from the coast road, known to the Allied defenders as

East Hill and the Charlies.

The guns that were available – a dozen Bofors, were dug in around the periphery of the runways, nine 100-mm and four 75s were well sited; the field pieces beneath the slopes of the Charlies. A Matilda tank was present on each flank of the airfield and the light Vickers were grouped below East Hill to spearhead any counter-attack. Like his opposite number at Rethymnon, Chappel was not inclined to allow the enemy any foothold. The Brigadier also ordered that, however tempting the target, his men were not to give away their positions by opening fire until the enemy were actually on the ground and thus at their most dispersed and vulnerable.

For reasons which remain unclear Freyberg did not seem to have any great confidence in Chappel, even though he was an experienced regular officer. He perhaps lacked the charisma of someone like Kippenberger but he was to give a good account of himself in the ensuing action. Heraklion was the most easterly of the main Allied concentrations, some 75 kilometres from Rethymnon and this meant that the brigadier's command was almost an independent fiefdom. Like the Creforce base, 14 Brigade HQ was tucked into a handy quarry, with Patrick Leigh Fermor attached as an intelligence officer. He had previously passed a pleasant three days at Prince Peter's comfortable coastal billet north of Galatas in the company of the Prince and Captain Michael Forrester of the Buffs who, like Leigh Fermor, had been attached to the military mission in Greece. Forrester, as piratical and swashbuckling as his companion, had been liaison officer for Brigadier A.G. Salisbury-Jones, reporting on the needs and readiness of the Greek units.

The two friends were reunited at Heraklion when Forrester inspected the local forces holding the town. They decided to dine together on the evening of 18 May; it would be the last such civilised recreation for some time. A third charismatic irregular present at this time was Pendlebury himself, fast becoming a legend amongst the fighters he was recruiting; a man whose empathy with, and understanding of, the Cretans surpassed generally by a very wide margin, that of his fellow officers.

For the airborne assault Major Bruno Brauer had three battalions of the 1 Parachute Regiment under his command, supported by engineers, artillery, machine gun, anti-tank and medical units. Despite the ferocity of the fighting unleashed by the morning's drops to the west, the defenders at Heraklion were blissfully unaware that an invasion was in progress. Their day so far was sufficiently leisurely that a number of officers had ventured into the town in search of a hot bath.

Around 4.00 p.m. however, by which time news of the other attacks had filtered through, the Stukas appeared along with Me109s, in sufficient numbers to clearly indicate that they presaged more than the daily aerial harassment. Despite the ferocity of the bombing the defenders' positions were sufficiently well hidden so as to avoid any casualties. Though the psychological effect of being shot up from the air is nerve-racking, troops well dug in amid mountainous terrain generally escaped serious loss.

An hour later the first of the transports began to roar in from the north-east along the coast. Immediately the Bofors guns opened up, the raucous concert joined by naval guns and Oerlikons manned by the marines. The slaughter amongst the slow moving juggernauts was great, the languid afternoon air rent by the quick firing clatter of the guns blowing chunks off the Ju52s and decimating their passengers. As many as fifteen planes came down in flames and another 200 paratroops were accounted for as they either sought to jump clear or scrabbled for their weapons on the ground.

One of the gunners, furiously hammering the German armada, recalled:

[I] saw planes burst into flames, then the crew inside feverishly leaping out like plums spilled from a burst bag. Some were burning as they jumped to earth. I saw one aircraft flying out to sea with six men trailing from it in the chord of their 'chutes. The 'chutes had become entangled with the fuselage. The pilot was bucketing the plane about in an effort to dislodge them.[2]

The attempt at a two-stage invasion had already placed a severe strain on men and machines; many aircraft had been lost, more were damaged. Added to this were those which crash-landed on return, got bogged in the mire and dust of the mainland airstrips or simply broke down. Numbers of troops had therefore simply been left behind and others were now dropped at extended intervals. At Heraklion a portion of Brauer's 2nd Battalion (Burckhardt) was dropped from under 200 metres directly onto the bayonets of the Highlanders on East Hill. Virtually all the officers became casualties, including the commander, Captain Dunz.

An eyewitness remembered one luckless parachutist drifting toward a cornfield where no less than eight British arose with fixed bayonets and dispatched their screaming victim like the English billmen of the Hundred Years War.

If this were not hardship enough the remainder of the battalion came down onto an open meadow, known as Buttercup Field, just west of the aerodrome and precisely into the sights of the Australians. A massacre ensued as the defenders opened up and the light tanks charged into the struggling knots of Germans who were shot down or crushed in droves. Only a handful of survivors escaped and these were obliged to take to the water and swim around to link up with the other survivors, grouped under Captain Burckhardt, on East Beach.

Brauer jumped to the east with 1st Battalion – too far east to offer any support to their comrades being cut down by the defenders' fire around the airfield. The drop was scattered over distance and time, darkness had fallen by the time Brauer was able to round up the majority. As the 1st Battalion advanced westwards, expecting to link up with the 2nd, they were stopped dead by well directed fire from the 2nd Black Watch. The darkness did however shield the infiltration of Count Blucher's platoon[3] which slid around the swell of East Hill to come within reach of the landing strip.

To the west the landing of the 3rd Battalion (Schulz) had also been widely dispersed, though they did succeed in putting in an attack against the walled town where the Greek garrison stood fast. Despite his fearful losses Brauer[4] was not the man to abandon the attempt. He ordered Schulz to desist from attacking the town and move in from the west (this order appears never to have been received).

Faced with the resolute defence of East Hill by the Black Watch, Brauer sent his companies into the attack piecemeal, convinced the effects of delay would be more serious than the lack of air cover available during daylight. This bold approach was thwarted by the Highlanders, who soon saw off their assailants, and obliged Brauer to withdraw to some high ground to the east, there to regroup and await the Stukas. Meanwhile Chappel sent his Matildas against Blucher's intruders, who were cut down to a man.

Schulz had received a signal from the Luftwaffe to the effect that the town was to be bombed next morning, the 21st, and so hung on, delaying a renewed assault until the defenders had been 'softened up' by the aerial bombardment. Despite the weight of bombs, the siren scream of the Stukas and the murderous hail of machine-gun fire, Schulz, advancing his battalions in twin columns, found the defenders still full of fight.

Becker's column was to strike through the North Gate with the objective of seizing the harbour. Despite stiff resistance the paratroopers blasted their way into the streets and succeeded in capturing the Old Venetian Fort guarding the harbour. The second column, under Egger, went in through the West Gate where the armed citizens gave them a warm reception; so hot was the welcome, the locals armed with captured German arms, that Egger's survivors were soon pinned up against the fort taken by their comrades.

German sources claim that a representative of the garrison offered to surrender at this point but that the British arrived in numbers to stiffen the defence. In fact this reinforcement consisted of nothing more than two platoons, drawn respectively from the Leicesters and the York & Lancs. Depleted, exhausted and with their ammunition all but spent, the surviving paratroops slipped away that night, back to their start lines.

To all intents and purposes this marked the end of the active stage of the Battle for Heraklion – the German assault had failed on all counts. The Luftwaffe continued to pound the town twice daily and the *Fallschirmjäger* vented their impotent fury on all those suspected of partisan activity in the areas they occupied; poor compensation for such meagre gains.

One of those to feel the fury of German frustrations was John Pendlebury. On 20 May he enjoyed a customary lunchtime drink in the basement bar of the Knossos Hotel. Nearby a bomb had blasted a supply truck and its cargo of fresh eggs had been miraculously converted into the biggest omelette in Europe. Even the most dismissive of his contemporaries, and there were indeed many ready to denigrate the fighting quality of the Greek nation, wrongly so, would have had to concede that Pendlebury had created the nucleus of a most efficient guerrilla network. He was, like Leigh Fermor and Forrester, a hugely romantic figure in the tradition of Byron, Sir Richard Burton or T.E. Lawrence. His partisans, men like the impressively moustachioed Manoli Babdouvas, the lean, hawk-faced Petrakegeorgis and the soon to be legendary Antonis Grigorakis, better known as 'Satanas', were formidable guerrillas. The history and topography of Crete, coupled with the fierce traditions of its proud people, would always ensure resistance to the invader. The casual German assessments that the locals would be, at worst, passive, were swiftly to be confounded.

Mike Cumberlege had been dispatched in his armed caique, now HMS *Dolphin*, to the eastern tip of the island, the Venetian port of

Ierapetra, to see if any arms could be recovered from a hulk sunk in the harbour. With the irregular's genius for improvisation, Cumberlege recruited a diver from the local Greek garrison. Pendlebury had given the crew a send-off dinner in Heraklion where the buccaneers discussed a possible raid on the Italian held island of Kasos, scheduled for the night of 20 May. Events, however, would move to frustrate this pleasant prospect.

With both of his other junior officers, C.J. Hanson and T. Bruce-Mitford dispatched into the mountains to mobilise resistance, Pendlebury had abided by his orders to remain in the city. His movements, during that fateful day, are unclear but it would have been entirely out of character had he not been involved in the fight to hold the ancient walls. The next day he made a characteristic attempt to slip away into the surrounding hills and reach his beloved partisans.

In the confused fighting and withdrawal of the surviving para-chutists, Pendlebury, armed and in uniform, accompanied by his driver and a squad of armed locals, drove through the Canea Gate. Having then parted from the escorting andartes he pushed on toward Chania. This was reckless, even for so noted a swashbuckler, as the area was thick with enemy. At Kamina, barely a kilometre further on, the vehicle ran into a blocking patrol and a fierce firefight broke out. Pendlebury, having accounted for a number of his opponents ,was severely wounded.

His captors, inclined towards mercy and carried the wounded man to a nearby house where two Cretan women administered first aid. That evening, he was attended by a German doctor who treated his wounds. On the following day, however, more German soldiers arrived, dragged the helpless officer outside and shot him on the spot. It is possible that, having been recognised as an SOE officer, he was singled out for execution.

Given the weakness of the Axis intelligence this is perhaps unlikely. More plausible is a further version which insists Pendlebury had been wearing a civilian shirt – as the doctor had, quite properly cut away the uniform tunic and the second party of Germans mistook their captive for a saboteur. Given the prevailing terror being experi-enced by the invaders, harried by soldiers and locals alike, this may offer a more realistic perspective.

Pendlebury's death was not the end of civilian resistance; quite the reverse. His killing is, however, clear evidence of the level of savagery the Germans would unhesitatingly employ against anyone suspected of partisan activity. The code of the paratrooper required chivalrous treatment of a uniformed enemy but no quarter to those who came from the shadows (see Appendix 2).

Meanwhile the Australians, in this sector, had captured a quantity of German flares and, having broken their signals code, were able to confuse the planes trying to bomb and strafe their positions, besides which they buried some 1,300 dead. Their own losses had not topped fifty. But that was it. The large garrison, having beaten off initial attacks, did nothing more, there was no determined onslaught to drive the survivors of Brauer's drop into the sea. Creforce HQ seems to have overlooked the potential of these substantial and victorious forces.

Without contrary orders Brigadier Chappel remained true to his original instructions which were to defend the town and airfield. An opportunity to drive westwards toward Rethymnon and link up with Campbell was therefore lost. It was not until the night of 26th /27th that Chappel approached his commander-in-chief, through signals routed through Cairo, to enquire if he should make aggressive moves in that direction.

By then, of course, it was far too late.

In Athens, any early optimism of the part of General Student and his officers was quickly dissipated. By 2.00 p.m. he knew the attack on Chania had been seen off with loss. The news from Maleme was equally dispiriting; using an ad hoc radio, cobbled together from spares and salvage, the paratroops reported more heavy losses and very limited gains. As a consequence the General wisely took the decision to postpone the afternoon's assaults on Rethymnon and Heraklion. These orders, nonetheless, did not all get through to the dustbowl airstrips, where confusion already reigned.

The choking dust in fact accounted for as many German transports as the ack ack fire they'd just come through. The planes were forced to land virtually blind in the great engulfing cloud, or circle for hours with their tanks running on empty. The net effect was that the bombers and fighter escort, supporting the second wave, took off in good time but the lumbering transports were significantly delayed. This provided a golden opportunity for the defenders to prepare an appropriate reception. The Germans' own report gave a bleak assessment of the results:

The formations started in incorrect tactical sequence, and arrived over their target areas between 1600 and 1800 hours, not closed up but in successive formations and at the most by squadrons. As the fighters, for reasons of range, could only remain over the landing targets till 1615 hours the bulk of the forces had to land without fighter protection. The bomber attacks started shortly before 1500 hours and had not destroyed the enemy but only kept him under cover for the time being. As a result of casualties to aircraft in connection with the first sortie the total battle strength at Heraklion alone was 600 men less than was intended.[5]

Major Rheinhard Wenning, who had taken off at the correct time expected, as he flew over the island, to meet the first wave as they turned for the mainland:

If everything had gone as planned we should have met the first planes on the way back but there was nothing ... a right turn then we fly parallel to the coast, signal the drop and the plane jerks as the parachutists jump out and float downwards. As far as I can see from the cabin of the plane there are no paratroopers on the ground even though those we have just dropped are supposed to be a reserve battalion. Now they will be facing an enemy superior in numbers on their own.[6]

It was only on his return journey that the Major observed the aircraft, which should have preceded him, approaching the coast! The 'fog' of war is not merely a literary phrase, it is certainly real and the more complex the operation, the greater the margin for confusion. Student's plan was visionary in concept, hugely ambitious in scale and marred by bad intelligence and, to a lesser extent, the petty jealousies of colleagues. Military undertakings are inevitably always fraught with difficulties; the plan subject to so many factors beyond the control of its creator:

It is not easy for the critic, strolling along an English country road in the pleasant and quiet darkness of a spring night to understand how difficult is to move bodies of troops even a few miles on a night of battle.[7]

By the evening of this long and bloody first day, General Freyburg was inclined to send a cautiously optimistic signal detailing the consequences of the day's actions:

Today has been a hard one. We have been hard pressed. So far, I believe, we hold aerodromes at Retimo [Rethymnon], Heraklion and Maleme, and the two harbours. Margin by which we hold them is a bare one, and it would be wrong of me to paint an optimistic picture. Fighting has been heavy and we have killed large numbers of Germans. Communications are most difficult. Scale of air attacks upon Canea [Chania] has been severe. Everybody here realises how vital the issue and we will fight it out.[8]

What is unclear at this stage is whether the General thought that the main thrust was still to come and that it would arrive by sea. The strategy of defence was totally hamstrung by this perceived imperative. Freyburg understood, quite clearly, that the airfields were important, even vital, but did he yet realise that they, and particularly Maleme, were to be the fulcrum upon which the outcome of the entire campaign would hinge?

The German view was less sanguine. The attackers, bloodied, depleted, scattered and stunned, were clinging to their broken hillsides, awaiting what they considered the inevitable series of counter-attacks with the clear conviction that these would quickly prove decisive: 'We should have had to fight them off with stones and sheath knives.'[9]

We have seen that the isolated companies of Andrew's 22nd New Zealand Battalion were coming under increasing pressure from the Germans who had landed, largely unscathed, west of the Tavronitis. As the long, hot afternoon wore on these had consolidated their gains, meagre as they appeared, and brought up their heavier weapons.

The Luftwaffe maintained the relentless aerial bombardment of Andrew's forward positions. The situation was not critical but it was deteriorating. Having secured the iron bridge over the dry river bed and established a bare toehold on the perimeter of the airfield, the Germans could yet gain control. This would be a disaster.

Whilst considering the chain of events which arose during the evening and night of 20 May and criticising the decisions taken, we must have an insight into the intelligence gathering which was informing Freyberg's reactions. ULTRA was Britain's great trump, to be guarded at all costs (see Appendix 3 The Intelligence War).

Freyberg himself was not a party to the exact source of the intelligence. As a result of this and also as a consequence of the manner in which the core data, the decrypts, were processed, the picture he was given, though highly accurate in detail, was also misleading. Freyberg was of the opinion that the airborne landings were merely the advance guard; the harbingers of a secondary but larger seaborne invasion.

The signals suggested this, as did military logic; Crete was an island, past invasions came by sea. The paratroop landings in the Low Countries in 1940 had preceded a sustained ground assault by conventional forces. As the Luftwaffe controlled the skies they had no need to fear the RAF, and their abundant air power effectively neutralised Britain's naval supremacy.

Freyberg, a soldier's soldier, blunt, brave, resolute, still thought, as did some of his officers, especially those who had served on the Western Front in the Great War, in terms of Great War tactics, linear battles of attrition where giving ground was less vital than conserving unit strength. They did not think of strategy in terms of revolutionary concepts such as vertical envelopment – indeed why should they? Such a battle had not so far been attempted and the ULTRA intelligence had been read to imply that the main thrust would come from the sea.

There is the unresolved matter of the extent of Freyberg's inclusion in the tight knit circle that guarded ULTRA. That such secrecy was vital cannot, of course, be denied. ULTRA was Britain's trump. Jumbo Wilson, during the retreat through Greece, was able to utilise the intelligence to consistently evade the threat of encirclement.

When the first serious indication of German numbers to be thrown into the invasion, was received on 6 May (see appendix 3), this suggested a force comprising two airborne divisions with supporting elements. So far so good, but what then muddled the issue was the German decision to retain a portion of the parachute troops in the Balkans and send in the Alpine division under Ringel.

A subsequent ULTRA signal instructed that: 'Flak units further troops and supplies mentioned our 2167 are to proceed by sea to Crete. Also three mountain regiments thought more likely than third mountain regiment'. This fresh intelligence appears to have confused London into believing that three mountain regiments were now to be deployed and that, crucially, these would come by sea.

Wavell was sceptical that the Axis could scrape together sufficient ships to create such an armada. A subsequent signal, dispatched next day from the Air Ministry,[10] gave a far more accurate analysis and stated the maritime element would be of only minor significance. Yet, on 13 May, yet another signal[11] reverted to the importance of the threat from the sea and, as a consequence, Freyberg was advised that he could expect to face 30, 000 or 35,000 enemy soldiers. Of these, 12,000 would be paratroops and at least 10,000 would be landed from ships.

The General's great strengths were in his steadfast courage and ability to lead men; intellectually and in terms of his analytical capacity he was less well endowed. In fairness ULTRA was both new and secret. His son, Lord Freyberg, has suggested that it was only relatively late in the day that his father was included within the ULTRA cabal. As a result he did not move troops to the airfields as this would have compromised the source.

This does not ring true. It seems far more likely that Freyberg was confused and misled. His obsession with an attack from the sea is palpable throughout and he does not appear to have awoken to the threat from the German possession of Maleme until 22nd, by which time the grip was too tight to be broken.

Churchill, usually quick to condemn, never lost his respect and liking for the battle-scarred hero, though recognising that Freyberg's type of warrior was out of his depth in the fine web of a modern intelligence-led war:

> Freyberg was undaunted. He did not readily believe the scale of air attack would be so gigantic. His fear was of powerful organised invasion from the sea. This we hoped the navy would prevent in spite of our air weaknesses.[12]

As the General himself subsequently conceded: 'We for our part were mostly preoccupied by seaborne landings, not by the threat of air landings.'[13] It was this misunderstanding and Freyberg's lack of analytical ability that doomed the defenders. On 20 May he had the upper hand, even if he did not fully understand the depth of the Germans' weakness. Determined action, ruthless and vigorous counter-attacks would have finished the battle on that day.

There is a supreme irony in this intelligence failure. On the one hand an army which has a complete knowledge of its enemy's intentions, an enemy whose own assessments are grossly inaccurate, misreads this gift to the extent that it saves the unprepared enemy from disaster. On the other hand, in the longer term, the defenders' failure helps to preserve the all important secret. As the German after battle report[14] concludes:

> One thing stands out from all the information gleaned from the

enemy (prisoner's statements, diaries and captured documents) that they were on the whole very well informed about German intentions, thanks to an excellent espionage network, but expected the bulk of the invasion forces would come by sea.[15]

It is perhaps significant that it was younger men, such as Kippenberger, who immediately understood the nature of the battle that was being fought and the means to win it. Counter-attack relentlessly, sacrifice what needs to be sacrificed to deny the enemy his prime objective which was to capture a viable airstrip intact.

Once the decision to retreat from Hill 107 was taken, the potential for final defeat was established, once Student's battalions, however mauled and depleted, took and could hold a single airstrip, the question of a seaborne attack became irrelevant. Once German reinforcements could be flown in then the balance swung in their favour.

Though the garrison, with a total strength of 42,460, outnumbered the total of the attackers at 22,040[16] by nearly 2 to 1, numbers would cease to count once the Germans began to build up reserves and heavy weapons. From a secure base in the west, and facilitated by air superiority, they could begin to roll up the defenders from west to east.

There were serious and ultimately fatal errors of judgement made by Allied commanders, particularly at the higher brigade and divisional levels, but these were failures of perception rather than nerve. The men concerned were never faint-hearted, quite the reverse; extreme bravery was a characteristic shared by all of them. It was strategic insight that was lacking.

Andrew's forward companies would hang on in the expectation that, once the sheltering dusk excluded the further attentions of the Luftwaffe, they could expect that a strong counter-attack might be put in. This was entirely reasonable for Allen's 21st and Leckie's 23rd Battalion had previously been ordered to hold themselves in readiness for just such a deployment.

The beleaguered Andrew was regularly sending up distress flares but these seem to have gone unnoticed. Major Thomason, at 23rd battalion HQ did not recall seeing any. Both worried and frustrated, Andrew decided he had no option but to make his representations in person. He was normally a calm and utterly imperturbable officer but, as Thomason records:

Colonel Andrew came to our Battalion headquarters in person

and asked for help. He was at first guided to me by one of our
men, I knew Les Andrew well, he and I were good friends. He
was very shaken and disturbed and I personally took him down
to Battalion headquarters. I don't know the outcome of his visit
except that his request was not granted. This took place fairly
early in the afternoon. I cannot state the exact time.[17]

This refusal was in spite of the fact a platoon sized patrol from the
21st had reported the Germans had a strong lodgement on Andrew's
southern flank at Vlakherontissa. Yet nobody moved. Both battalions concentrated on holding their own positions while reporting
back to brigade that all was quiet in their respective sectors.

Brigadier Hargest confirmed that nothing further would be
required of them 'unless position very serious'.[18] In fact the position
was quickly becoming very serious indeed and Andrew's[19] messages
to brigade were anything but relaxed. He reported that he'd lost
contact with his forward companies, dug in on the western flank of
Hill 107, that his positions were being bombed relentlessly and that
the German strength was growing with a corresponding increase in
their deployment of heavier weapons.

These messages became increasingly desperate as the day progressed.   Andrew reported that his HQ company had been
penetrated, that the left flank of his battalion was giving way. At
5.00 p.m. he bluntly demanded to know when he might expect reinforcement; when would the 23rd appear?

Hargest did not immediately respond but when he did it was to
advise that the 23rd was already fully committed – this was entirely
fallacious.   Having seen off Scherber's attack the battalion had
scarcely fired a shot. Inevitably Brigadier Hargest's conduct has been
much criticised. However, such criticism must always be considered
in the light of the prevailing fears of an invasion from the sea.
Paratroop landings, in this context, were no more than either a
diversion or an overture. Hargest was certainly a very brave man, an
eyewitness at Thermopylae remembered the General:

> ... standing in the centre of a group among some olive trees and
> saying, with tears running down his cheeks, 'I have in the past
> had to retreat when I have been driven out of a position, but this
> is the first time I've had to retreat without a fight'. My second
> picture of him is at Sidi Azziz on the morning of 27th November
> 1941. We had been attacked by a force of German tanks, artillery

and motorised infantry and the guns of E Troop, 5th Field
regiment had been knocked out one by one. Before that the anti-
tank and anti-aircraft guns had been knocked out and we were
obviously losing the battle ... Hargest stood a little to the right
of the line of fire of our 25 pounders, and in front of them,
wearing his red cap and with his hands in the pocket of his coat.
He did not take cover at any time. We were firing over open
sights at tanks, and when the last of our guns was out of action
and our ammunition truck caught fire, Hargest put his hands up
and walked towards the German tanks.[20]

Whatever the reason for this extraordinary reply, Andrew felt that all
he could now do was to mount a local counter-attack with whatever
he had in hand. North of the beleaguered Hill 107 the Colonel had
stationed his pair of Matildas which were now to be thrown into the
attack in an attempt to drive off the Germans on the periphery of the
airstrip and retake the bridge over the Tavronitis. One of the
strategic failings surrounding the Battle for Crete, as some commen-
tators have pointed out, was the dispersal of the available armour.
Using tanks in penny packets to support local infantry attacks had
been a weakness in the Battle for France – it remained so here on
Crete.

The actual usefulness of heavy tanks such as the Matilda for operations in such harsh and hostile terrain was also subsequently
scrutinised. The After Battle Report concluded that the light tanks
were very effective in dealing with enemy machine guns and nests of
snipers.

Safe from small-arms fire they could hose the ground with their
own machine guns then charge home as light cavalry, crushing or
overrunning the foe; any survivors could be dealt with on a hand to
hand basis. If these tactics sound exceedingly brutal, they were
certainly effective. Whether the heavier infantry tanks could perform
as well in this close support role was far less certain.

Andrew dispatched Peter Butler, formerly of C Company, as a
runner in mid-afternoon, to establish contact with Captain Johnston.
On his way, hastened by threatening bursts from MG 34s, he
blundered into a well-concealed tank. The commander advised him
that he was still awaiting orders before engaging. Butler spent over
an hour crouched with his former comrades in their exposed
positions before being sent back with a situation report. Returning
around five, he was in time to witness the counter-attack, spear-

headed by the Matildas.

The line of attack had been worked out between Johnston and the tank crews the day before and, as the Captain records:

> The tanks left their concealed position at 5.15 and moved west past company headquarters along the road towards the river in single file about thirty yards apart ... the second tank turned about before reaching the bridge and came back past company headquarters on the Maleme road.[21]

This inspiring beginning proved anti-climactic for, as Butler records, the second tank soon came lumbering back. Apparently the ammunition for the machine gun was of the wrong calibre and the traverse mechanism for the turret was jammed. The first of the tanks, however, managed to descend into the dry river bed from the road and move forward under the span of the iron bridge. The ground was alive with Germans who were badly rattled by its appearance.

Student later admitted that his men, already dazed by their heavy losses, were reduced to something akin to panic by the arrival of British armour, which was impervious to small-arms fire. Gericke, amongst those crouched in the rocks, confessed that this new development caused ripples of fear amongst his tired men. However, the threat proved more apparent than real. The broken ground afforded plenty of cover and the tank was not, in fact, able to inflict many casualties.

Matters quickly degenerated for the British. This tank soon stalled and, with its turret also jammed, the crew baled out and were captured. The lead platoon from C Company, following in the wake of the metal monster, was caught in a most unfavourable position, assailed from all sides. The survivors were forced into an ignominious and costly retreat, leaving a scattering of dead and wounded. Of the attacking sections only three men came back unwounded.[22] A gallant gunner who had, with his comrades, volunteered to fight as a foot soldier when their gun had been knocked out earlier, was amongst those who failed to return. The whole sorry business was over in half an hour and with nothing to show for the loss.

It would be easy to find fault with Andrew for launching a small, local counter-attack which could only bring very limited resources to bear against an enemy now well dug in and over very difficult ground. This would, however, be unfair. The Colonel was acutely aware of how exposed his forward companies were and of the need to relieve them from the relentless pressure now building up from west of the Tavronitis. For Andrew the failure to establish defences in this vulnerable sector was now bearing bitter fruit. His oft repeated pleas for additional support had gone unheard; it is difficult to see what other choices were available to him.

With his attack stalled and fearing the two forward companies had been overrun, Andrew spoke again to Hargest and warned he might have to withdraw from Hill 107. The Brigadier's reply was 'If you must, you must'.[23] In fact both Johnson and Campbell were hanging on. A and B companies had barely engaged, Andrew nonetheless proposed to fall back to B Company Ridge, a spur that lay below Hill 107 and was, in fact, very much in its shadow.

Despite the desperate fighting in Andrew's sector and the clear implications of abandoning Hill 107, Hargest does not seem to have felt the need to inspect his forward battalion, preferring to remain at Brigade HQ. The question therefore arises as to whether he felt that he had to conserve his strength to repel a second, seaborne attack on his sector. If so, and it may well be the case, then ULTRA had, in fact, worked for the Germans rather than the Allies by causing them to 'look the wrong way'!

Having thought again, however, Hargest decided to detach two companies – one from 23rd Battalion and a company of Maoris from the 28th, despite the fact they were a good eight miles distant. Andrew, being apprised of this then, for whatever reason, continued with his phased withdrawal from the hilltop and from the eastern perimeter of the airfield. By 9.00 p.m. the relief company from the 23rd was to hand and Andrew sent them off toward the rise of Hill 107 from which his own men had just withdrawn.

The Maoris were held up, became lost and ended up in a satisfying but pointless firefight with a group of parachutists barricaded in houses by the coast road. The obstacle was cleared and prisoners taken but Andrew, physically and emotionally exhausted, had decided to withdraw his whole force toward 21st and 23rd Battalions, abandoning Hill 107 again and take position on B Company ridge. With this was lost the infinitely greater prize of the airfield itself and the situation lurched from deadlock toward disaster.

Suffering from heavy casualties and having been profligate with their ammunition, the Germans were scarcely placed to renew the attack. C and D companies, despite losses, were by no means beaten; 'still full of fight' though their ammunition was lower than their spirit.[24] Campbell, who was expecting his position to be utilised as a jumping off for a fresh counter-attack, at first refused to believe that he had been abandoned. Only when he and his CSM confirmed, from personal reconnaissance, that Battalion HQ was an empty shell, did the full realisation sink in.[25]

The order to split into three detachments and filter southward through the tangle of hills must have been a galling one, to give up a position so bravely held and break off from a fight that seemed to be winnable. Johnson, even more exposed on the seaward side of the aerodrome, hung on till first light then, unable to establish contact with anyone from battalion, led his men back toward the 21st. The exhausted Germans simply let them go but it was a tragic end to so determined a stand.

An opportunity, not one to be repeated, had now been lost and the debate over who was to blame has raged ever since. Freyberg, as C.-in-C., had always regarded Maleme as the prime Axis target. Captured orders have since confirmed this, so the degree of laissez-faire which obtained at Brigade and Divisional HQ seems hard to comprehend. One explanation, which has been offered, is that Hargest was in the habit of delegating visits to the forward positions to his immediate subordinates. There is, of course, in theory, nothing wrong with this. The brigadier should, in a conventional battle, remain so placed that he can respond to events across the board.

This, however, was not a normal battle. It was not a conflict of attrition or even of mass manoeuvre – it was a fight for a single key objective, which would unlock not just the sector but the entire Allied position. It has been suggested that:

> Had Brigadier Hargest gone to his forward Battalions himself instead of sending his Brigade Major (Captain Dawson) there might have been a different story to tell. Surely he would have vetoed the withdrawal of 22nd Battalion from the airfield. Surely he would have launched a counter-attack, and his presence would have inspired the troops at a time when inspiration was needed.[26]

1. British, Australian and New Zealand troops disembark at Suda Bay, Crete.
(Alexander Turnbull Library, Wellington, New Zealand. Photo DA-01611)

2. Alpine troops boarding Greek vessels.   *(Courtesy of the Naval Museum, Chania)*

3. General Thomas Blamey, in command of Australian forces in the Middle East.
(Courtesy of the Naval Museum, Chania)

4. Airborne parachutists leaving planes to land on the island of Crete, below.
(Alexander Turnbull Library, Wellington, New Zealand. Photo DA-08215)

5. Dead German paratrooper enshrouded in his parachute.
(Courtesy of the Naval Museum, Chania)

6. German paratroops in a street in Canea, as prisoners of war, captured by 2 NZEF in Crete. *(Alexander Turnbull Library, Wellington, New Zealand. Photo DA-01153)*

7. The grim task of burying German dead. *(Courtesy of the Naval Museum, Chania)*

8. A German troop carrier on fire over Crete during the Second World War.
(Alexander Turnbull Library, Wellington, New Zealand. Photo DA-07491)

9. Soldiers marching to transit camp in Crete. Officers at head of column are Major
R.L.C. Grant *(left)* and Captain D.W. Burns, Divisional Signals.
(Alexander Turnbull Library, Wellington, New Zealand. Photo DA-10468)

10. German parachutes near Galatos hung up in olive trees. Shows soldiers Cyril Ericson (holding map case) and Ernie Avon.
(Alexander Turnbull Library, Wellington, New Zealand. Photo DA-00470)

11. New Zealand soldiers keeping under cover in a cave while awaiting evacuation from Crete. *(Alexander Turnbull Library, Wellington, New Zealand. Photo DA-10636)*

12. The iron bridge over the dried up bed of the Tavronitis. Failure to dispose Allied troops west of the Tavronitis to meet the airborne threat prior to 20 May was a significant factor in permitting Student's badly mauled paratroops to gain an unchallenged foothold west of the dried up river bed. *(Author)*

13. Looking back (northwards) up the Imbros Pass, the modern road leads down to the south coast in a dizzying sequence of tight hairpins but in 1941 the road was unfinished and the retreating troops were faced with a rough scramble down the forbidding reaches of the gorge itself. *(Author)*

14. Memorial Cross at the German war cemetery Maleme. *(Author)*

15. The war memorial, Galatas. *(Author)*

CHAPTER 6

The AA Waltz—Battle at Sea

Before the Greek invasion Admiral Cunningham made his daring and successful sweeps through the Mediterranean in complete disregard of the Italian Air Force. When Italian fliers did attack, the fleet put up an AA barrage [and] the Italians took avoiding action ... But over Crete Waters, German pilots came out of the sun in steep power dives, utterly disregarded AA fire [and] released their bombs over the target ... dive bombing was accompanied by high level bombing and torpedo attacks. Often the bombs struck before the bomber was seen. The fleet AA could only fire barrages into the sun [and] hope fir hits ... In some cases of major damage or sinking the air attack had been of such intensity and duration ... that the vessels were out of ammunition long before the bombing ceased.[1]

I had the vessel [in my bombsight] ... From bows to stern she filled the circle, and then with decreasing distance she seemed to grow fast ... this was a cruiser and now I saw two more of them in line ahead. This was something I had never seen before ... My cruiser ... shot at me with every gun barrel and her speed was so fast that she forced me to flatten my dive ... I pushed the button immediately turning to starboard and the bombs dropped. I was now within easy reach of the light guns and the tracers ... were everywhere ... I would have given a fortune for more speed to get out of the range of those gunners. All of a sudden there were cascades of water coming up my way. They shot at me with heavy artillery planting water trees right in my course ... I began to dance ... 'The AA Waltz', turn and turn upwards and downwards ... It was not fun, however. I felt that there were professionals firing at me ... This was my first encounter with British cruisers, I said and I am still alive, still flying ... home to Eleusis.

The bombs had hit the wake.[2]

Some two centuries ago Admiral Lord Nelson had defeated the combined fleets of France and Spain off Cape Trafalgar, securing Britain from invasion and establishing the fact of British naval hegemony. 'Britannia Rules the Waves' was not mere jingoism, it was a factual statement and no maritime power came close to challenging the global supremacy of the Royal Navy until the Naval Arms Race with the German High Seas Fleet in the years prior to the outbreak of the First World War. Though the single clash between the opposing dreadnoughts at Jutland in 1916 was inconclusive, the High Seas fleet retired to harbour and made no significant sorties.

The development of submarine warfare and the increasing importance of air power had changed the face of war at sea. The loss of the French fleet after 1940 and the entry of the Italians, with their powerful navy, into the war threatened the whole British position in the Mediterranean. Beaten at Taranto and off Cape Matapan, the menace from the Italians largely evaporated but that posed by the Luftwaffe did not.

If the attempt to seize Crete by aerial envelopment represented a significant tactical innovation then the concurrent battle at sea would see a major air force take on a dominant battle fleet in a savage and sustained fight, the ships utterly unprotected from the air other than by the weight of their own AA barrage. The experience would, for the British Navy, be both chastening and costly.

By May 1941 Admiral Cunningham's ships carried an onerous burden, or series of burdens, shadowing the more numerous Italians, shielding Malta, blockading Libya, supplying the outlying bastion of Tobruk and, most recently, shipping the army to and rescuing it from Greece. All at a time when the RAF was least able to provide supporting air cover. Before the Battle for Crete began Cunningham was noting that his ships and crews were worn out, the vessels needed servicing and refitting and the men needed rest from the constant strain of action. Ammunition, particularly for the AA guns, was running at dangerously low levels.

During the course of the forced evacuation of the army from the small ports and beaches of the Peloponnese, the strain had increased significantly and twenty-two ships had been lost. None of these went down in ship-to-ship engagements; all were sunk by marauding German bombers and dive bombers and it was clear that the main danger came from the skies. The Italian ships, rarely sighted, now declined to engage and the Axis had no fleet other than theirs available, but the Luftwaffe more than compensated for the deficiency.

The cardinal lesson learnt from the Greek fiasco was that naval operations could now only proceed safely under cover of darkness; to be exposed at sea in daylight hours was to court disaster. Cunningham, understandably, had been wary of his ships' ability to protect Crete. One of the bitter lessons learnt, especially after the loss of the two destroyers *Wryneck* and *Diamond* off Nauplion, was that single ships or pairs were particularly vulnerable. Any foray needed to be squadron sized so the combined AA barrage could create a storm or 'box' of shot, forming an anti-aircraft umbrella.

As a seaborne descent on Crete seemed likely, repelling any such armada was Cunningham's prime strategic role. Consequently he deployed his ships in three battle squadrons. Vice Admiral Pridham-Whippel and Rear Admiral Rawlings were given charge of the main battle fleet, itself divided into forces A and A1 – their allotted task was to cruise the western approaches to intercept any attempt by the Italian Navy to cover a landing. This powerful force included several capital ships: *Queen Elizabeth*, *Barham*, *Warspite*, *Valiant*, two cruisers and sixteen destroyers.

To see off any intruders Cunningham created two cruiser squadrons. Rear Admiral King (Force C) commanded *Naiad*, *Perth*, *Calcutta* and *Carlisle*, with four destroyers whilst Force D, under Rear Admiral Glennie, comprised: *Dido*, *Orion*, *Ajax* and another four destroyers. To avoid the perils of daylight sailing, the two cruiser forces would stay to the south and then, under cover of darkness, sweep the north coast; King bearing west once through the Kaso Strait with Glennie passing through the Antikithera Channel and sweeping eastwards.

The 20 May saw no major action at sea but on the 21st, as the Mediterranean dawn poured light over the 'wine dark sea', King's squadron had a brush with the Luftwaffe which resulted in the loss of *Juno*.[3] As the day wore on the evidence of an attempted landing, gleaned from intelligence sources, mounted. Admiralty intelligence on Crete, managed by Captain J.A.V. Morse, was particularly efficient, relying on an established espionage network supported by aerial reconnaissance (sea planes, the Short Sunderlands, flying from Egypt). By midday Cunningham knew what was afoot and issued orders to King and Glennie to continue their coastal sweeps that night.

As the fast warships, moving at full speed, surged along the coast of Crete, the rather ramshackle fleet Student had been ordered by Hitler to assemble, comprised some twenty odd commandeered caiques, a handful of coastal steamers and only a single escort, the Italian destroyer *Lupo* and four torpedo boats; on board some 2,330 soldiers, mostly from Ringel's 5th Division, with an array of heavy weapons, field and AA guns, tanks, motorcycles and transport vehicles. For these Alpine troops the sea was as foreign as the skies: 'Few of us had been on board a ship in our lives.'[4]

At 11.15 p.m. Glennie's ships fell upon this flotilla in a singularly one sided engagement. The job was quite simply to kill Germans, as many as possible, and to send their heavy weapons to the bottom of the Mediterranean. For the men on board the British warships this was an opportunity to strike a blow – to make up for the hours of nerve shredding bombing and strafing they'd endured. It did not go to waste; for two and a half hours Student's armada was rammed, shot up and machine gunned; few of the leading vessels survived.

It was distasteful to fire on the hapless Greek crews press-ganged into hated service but this was war at its most brutal and business like. The job was to frustrate a seaborne attack and this was ruth-lessly accomplished. The German flotilla, as a tool for invasion, was a total write-off, neither men nor any matériel reached the shore. The Royal Navy had fulfilled its promise to the letter. Jahnke, a soldier with the 5th Division, recounted the terror of the ordeal:

Suddenly and without warning the sky was filled with brilliant white parachute flares which lit up whole areas of the sea. The blinding light lasted for about three minutes ... then searchlights swept across the water and fixed on the ship ahead and to our port side. We saw several flashes from behind one of the beams and soon realised that these must be from enemy guns because ... shells began to explode on that caique. Soon she was alight and we could see our boys jumping into the sea. Our ship was illu-minated by the fire and the lieutenant told us to put on our life jackets and to remove our heavy, nailed boots. Barely had we done this when we too were caught by the searchlights ... We formed two ranks. Our officer called 'Good luck boys!' and ordered our first rank to jump into the sea ... All this happened in less than five minutes ... [but] everything seemed to slow down so that it seemed as if hours had passed. The water was very cold and the shock of it took my breath away.[5]

The *Lupo* did her best to protect the convoy. Hopelessly outnumbered and outgunned, her captain fought his ship with great skill, courage and determination, despite the many hits she took. Having accomplished his mission, Glennie withdrew his squadron while still shrouded by the cloak of darkness. Cunningham was later to criticise this decision for it left King, still sweeping the north coast as dawn began to filter, exposed.

It was known that two flotillas had left the mainland. One was accounted for but that still left another which could not be allowed to proceed unmolested. Force C, therefore, continued on its northward course, soon sighted by enemy spotters and then harassed by scores of dive bombers. The AA Waltz struck up its deadly tune, Bofors and Oerlikons blasting in continuous, ear shattering concert throwing up the umbrella of fire and the enemy aircraft swooped and screamed like flocks of demonic gulls.

Mid morning and the British were less than a score of miles from the island of Milos, having already passed the small islands of Ios and Thera.[6] Then, having sighted a scattering of ships and a couple of enemy destroyers, the second convoy was discovered, a plum ripe for picking. At this point, however, King decided not to attack, a decision which later brought obvious criticism.

His expressed reasoning was that his ships' supply of ammunition was already too low; he then ordered them to discontinue the attack and clear away westwards. Cunningham, when he read his subordinate's signal to this effect, was outraged and immediately sent a reply ordering King to engage and destroy regardless of hazard. But by now the convoy had panicked into a random dispersal and scattered – the effect, in terms of the fight for Crete, was the same as it if had been sent to the bottom; neither troops nor equipment reached the island. The Navy had not broken its promise.

For the sailors of Force C there was another pressing concern, the prospect of a four hour dash to the Kithera Channel under continuous and sustained aerial bombardment. Now the might of the Luftwaffe, their aircraft barely thirty minutes' flying time away, was pitted against the skill and resolution of the Royal Navy and its exhausted gunners who faced the further, potential horror of their ammunition running out.

With *Carlisle*, the slowest ship incapable of more than twenty knots, and *Naiad* damaged by bombs, the duel began in earnest. The more damage sustained by individual ships and the slower the overall speed and manoeuvrability of the fleet, the quicker the precious ammunition was expended. Realising the desperate plight of his ships Cunningham ordered the main battle fleet, which now included Glennie's squadron, to steam eastwards and add their greater firepower to the fight.

It was now a fight to the finish. Like hungry hawks, spotting their prey below, the Luftwaffe pilots pounced, again and again, throwing their planes against the wall of fire, those refuelling, rushing back into the storm. *Warspite* was hit and damaged, losing part of her formidable arsenal. *Greyhound* on a solo mission was bombed into oblivion, sinking in fifteen minutes. *Kandahar* and *Kingston* were detailed to pick up survivors, supported by the guns of *Gloucester* and *Fiji* – both withdrawn when it was realised in what parlous state their ammunition reserves stood. This was an opportunity which the circling predators were sure not to miss, nor did they; *Gloucester* was lost.[7]

Charles Maddon, Executive Officer on *Warspite,* provided a graphic account of what it was like in the confined spaces of a warship when such colossal punishment struck:

> One four inch mounting had gone overboard completely .. there was a huge hole in the deck ... from which smoke and steam were pouring out. I ... went down to the port six inch battery ... to try to get at the seat of the fire through the armoured door that connected the port and starboard six inch battery decks ... We had great difficulty in opening the door and had to use a sledge-hammer. Finally, it gave, to display a gruesome scene. The starboard battery was full of flames and smoke, in among which the cries of burned and wounded men could be heard. This was very unnerving ... I was soon joined by more fire parties ... but was hampered by the continued cries of the burned men, which distracted the fire parties who wanted to leave their hoses to assist their comrades. I therefore concentrated on administering morphia ... As it was dark and wounded men were thrown in all directions amongst piles of iron work and rubbish this was not easy .. I then went to the starboard mess decks where a fresh scene of carnage greeted me ... When all was in control I went to the bridge to report. The calm blue afternoon seemed unreal after the dark and smelly carnage below.[8]

The gallant *Fiji* fought a heroic but doomed fight; by 8.15 p.m. she turned turtle and went down. At least the two destroyers were on hand to pick up survivors as the sheltering darkness once again enfolded the battered ships. Cunningham's last and freshest reserve was Mountbatten's 5th Destroyer Flotilla. This he committed to a further sweep of the north coast via the Kithera Channel. Due to an error in the signals which confused the words 'empty' and 'plenty' – a significant confusion when reporting upon the capital ships' supplies of ammunition – Cunningham had decided not to commit his ships to a further night's action.

This now left the five K class destroyers of Mountbatten's squadron exposed in their westward dash to the Channel. Sleek and very fast, these nimble vessels confounded the dawn chorus of dive bombers but, as the morning wore on, the attacks intensified. *Kashmir* was the first to succumb, soon followed by the flagship *Kelly* which continued her mad 30-knot dash even in her death throes. Had not one of the surviving vessels turned back to pick them up then all aboard would likely have been lost. Despite the fury of the onslaught unleashed against her, *Kipling* was successful in rescuing survivors and in returning, amazingly unscathed, though out of fuel, to Alexandria.

In the course of the sea and air battle the Navy had prevented the German convoys reaching the battle zone. Admiral Cunningham had been true to his promise, but the cost to the Navy in men and ships had been fearful. The survivors, limping into Alexandria, were in a frightful state. The Admiral thus intimated to London that further naval operations would be too costly to contemplate:

> I am afraid that in the coastal area we have to admit defeat and accept the fact that losses are too great to justify us in trying to prevent seaborne attacks on Crete. This is a melancholy conclusion but it must be faced.[9]

By way of reply the chiefs of staff (or more likely the Prime Minister) remained obdurate, stressing the need for operational sorties to be made, in daylight if necessary, in support of the land battle, regardless of the scale of loss that would, as it surely must, result.

Cunningham was brutally direct in his riposte:

> It is not the fear of sustaining losses which will cripple the fleet without any commensurate advantage which is the determining factor in operating in the Aegean ... The experience of three days in which two cruisers and four destroyers have been sunk, and one battleship, two cruisers and four destroyers severely

damaged shows what losses are likely to be. Sea control in the Eastern Mediterranean could not be retained after another such experience.[10]

In the first major trial of strength between a conventional battle fleet and a determined air force, the destructive capacity of the planes had been clearly demonstrated. Victory had gone to the Luftwaffe though the Navy had made good its promise to thwart any seaborne reinforcement. The question would now have to be resolved by the troops of both sides so hotly engaged on the ground.

CHAPTER 7

Bombed from the Earth—
the Turning Point

I should have realised that some of my Commanders, men from World War One, were too old ... to stand up to the strain of an all-out battle of the nature that eventually developed around Maleme Airfield ... I should have replaced that old age group with younger men who ... stood up much better to the physical and mental strain of a long and bitter series of battles.[1]

In the illusory baroque splendour of the Hotel Grand Bretagne, General Kurt Student was a very worried man. He had just cause. Though early reports, during the morning and early afternoon of 20 May, had provided an optimistic picture, more detailed assessments, coming in through the evening, painted an altogether different picture. The operation appeared, at this stage, a total shambles – a much more numerous and better prepared enemy, catastrophic losses and nothing, particularly not a single airstrip, to show for it all.

That evening the General's staff were joined by both Lohr and Ringel, neither of whom was overly well disposed toward Student. If the affair did turn into a fiasco there need be no question of where the full weight of the blame should land. Casualties had been heaviest amongst the officers; those troops who remained on the ground were scattered and, in many cases, leaderless.

An early hope that the strength of the reception at Maleme implied Rethymnon and Heraklion were only lightly held, had been dashed. Student knew that most of his officer colleagues could be counted as enemies. These would not hesitate to withdraw the battered survivors if the tactical situation did not improve dramatically and soon.

Lohr, like most senior officers in May 1941, was driven by the need to conform to the demands of Barbarossa. Ringel could scarcely be expected to willingly hazard his division in an operation which appeared to have already consumed most of 7th Airborne without establishing a viable bridgehead.

A lesser man than Student might have suffered a loss of nerve but he was prepared to cling to the vestige of hope that the situation at Maleme appeared to offer. At this point the airstrip was not taken but the fact the attackers had even the flimsiest of toeholds, justified the commitment the remaining few companies held in reserve:

> At no point [Student observed later] did we succeed completely in occupying an airfield. The greatest degree of progress was achieved on Maleme airfield, where the valuable assault Regiment fought against picked New Zealand troops. The night of May 20th/21st was critical for the German Command. I had to make a momentous decision. I decided to use the mass of the parachute reserves, still at my disposal for the final capture of Maleme airfield.[2]

Having faced down his opponents, at least for the moment, the General retired. He was under no illusions, nor did he sleep but stayed awake all night to await further news; his pistol on the bedside table. He had no doubt as to the course expected of him should the final failure of the attack have to be conceded; failure in the Third Reich was not permitted.

The General might indeed have been tempted to reach for his Luger had he been aware that the operational order for 3 Parachute Regiment was in Freyberg's hands. This document, besides the detail on individual unit targets, made it quite clear that the Germans were very considerably at risk from a single, concerted counter-attack.[3]

Crucially, as this vital intelligence was being digested, 5 Brigade's battalion commanders had convened a hasty conference at which Colonel Leckie was presiding. No decisions were taken other than to consolidate, despite the fact that they knew the enemy to have suffered serious loss whilst their own had been considerably less. Brigadier Hargest seemed no more aggressively inclined.

When the utterly exhausted Andrew arrived at his HQ, by Bren carrier at 5.00 a.m. on the 21st, Hargest merely advised that the 22nd Battalion must stay in the line but he made no comment on the need for counter-attack. When Andrew, with the Brigade Major, Captain Dawson, returned to his men, their joint orders went no further than 'discussing' new defensive arrangements.

The clock that would finally decide the fate of Crete and its garrison was already ticking and it is likely that Freyberg was correct in his retrospective analysis. The men leading 5 Brigade were brave and competent. Their failing was not dereliction of duty but one of comprehension. They did not understand the true nature of the battle that was being fought.

Student possessed one inestimable asset – he was in radio contact with the survivors at Maleme. He had little more than 500 fresh troops in hand. To gain a clearer view of the situation at Maleme he detailed Captain Kleye from his staff to embark on a personal reconnaissance. Kleye was an excellent choice:

> ... a bold go-getting character on my staff and told him to take a Ju52 and land at Maleme in order to get a personal feeling of how things were going with the Storm Regiment ... he managed to land on the airfield and also to get off again although fired at by the enemy. In this way he was able to bring back the important information that the western edge of the airstrip lay in dead ground.[4]

His aircraft touched down on the airstrip at Maleme around 7.00 a.m. on 21st and, despite intense ground fire, successfully unloaded its much appreciated load of ammunition. At virtually the same time a Lieutenant Koenitz, acting on his own initiative, managed to land another transport on the beach at the mouth of the Tavronitis. He too carried more precious munitions; many of the survivors were down almost to their final rounds.

Amongst the wounded Koenitz was able to evacuate, was Meindl himself, now delirious from his wounds. Thereafter reinforcements from the reserves began to arrive. Kleye returned to report that the day was by no means lost and that an opportunity existed to secure the airstrip. Despite the withdrawal of Andrew's companies the German assault on Hill 107 was not unopposed, indeed the Germans reported some heavy fighting, indicating they'd run into Cretan partisans and stragglers from 22nd Battalion.

Student's revised plan was that the remnants of the Assault Regiment, once their grip on Hill 107 was assured, should push along the coast toward Pirgos. Colonel Bernhard Ramcke would lead the fresh drop which would straddle the airfield east and west. The advance was slow and halting, the paratroopers still shaken by their rough reception the day before, calling down the Stukas whenever they caught a whiff of opposition.

By mid afternoon, however, they had successfully occupied both Maleme and Pirgos. The airstrip, though not yet beyond the range of the defenders' guns, was in their hands. The tide had begun to flow and, once turned, would be impossible to halt.

The New Zealand officers, other than sniping and some long range firing, did nothing to check this advance; rather they concentrated on consolidating a linear defence with the remnants of 22nd Battalion re-deployed in the gap between the other two. As the German patrols began to probe the line, and air attacks intensified, the defenders braced themselves and, as the Germans then attacked, opened a withering fire which accounted for as many as 200 of the paratroops.

Major Wenning was, once again, back in the air and directing the drop:

> As we reach the coast and turn towards the dropping-zone we can see the fighting below. Everywhere in the area we see yesterday's parachutes like countless points of light below. During the drop my 'plane suffers two mishaps. One parachutist, just preparing to jump, is badly wounded by a shot from the ground but he jumps anyway. And the last man to go baulks; he does not want to jump. We have already passed over the ground where we have put down his unit but he must jump and now he is ready to do so. So we repeat the manoeuvre while the other planes turn for home. And we fly out to sea, turn towards the coast, dropping to 150 metres off the ground. All the fire of the defence is now concentrated on our plane but despite this we succeed in reaching our position. The man jumps and now we turn and head at full speed toward the open sea. Flak riddles our plane but causes no serious damage.[5]

Barely had the sound of firing died away than the eastern prong of Ramcke's assault began its jump. In the mistaken belief that the New Zealand line ran through Pirgos and Maleme, these men, under Lieutenant Nagele, were dropped west of Platanias in an attempt to encircle the Allied survivors. Instead the parachutists experienced a repeat of the previous day's fighting, falling directly onto the bayonets of the defenders.

The Maoris leapt to their feet and, with their terrifying war cries, fell upon the invaders, most of whom were swiftly accounted for.

Nagele managed to round up eighty or so survivors and barricade some farm buildings. Churchill would have thoroughly approved and this type of fighting did indeed suit the New Zealanders, 'down to the ground'. As the Germans were descending, HQ platoon:

...formed up with two sections forward and advanced unopposed, a brisk exchange of grenades and small-arms fire resulted in ten dead Germans on the road and others in the scrub with no loss to the makeshift platoon .. and then back to Battalion Headquarters passing en route a number of dead paratroopers in front of C Company lines. Ngatipourou had made the most of such opportunities as had come their way.[6]

Captain Anderson reported that:

At one stage I stopped for a minute or two to see how things were going and a Hun dropped not ten feet away. I had my pistol in my hand and without really knowing what I was doing I let him have it while he was still on the ground. I had hardly got over the shock when another came down almost on top of me and I plugged him too while he was untangling himself. Not cricket I know, but there it is.[7]

The Maoris also attended to numbers of Ju52s landing on the stony shale of the beach, their Bren guns doing fearful damage, blowing great chunks from the thin fuselages and shredding the packed troops within, few of whom survived long enough to return fire. Those who made it clear immediately became casualties or prisoners. The fighting was brief but murderous:

... One at about 15 yards, instead of firing his Tommy gun started to lie down to fire. I took a snap shot with a German Mauser. It grazed his behind and missed between his legs. My back hair lifted but the Maori got him (I had no bayonet). We rushed on ... some tried to crawl away ... a giant of a man jumped up with his hands up like a gorilla and shouting 'Hants oop'. I said 'shoot the bastard' and the Maori shot him. That was because many others were firing at us and a Spandau from further off. Suddenly bullets spluttered all around my feet ... .[8]

Unaware of this fresh disaster Ramcke, with the western contingent, had dropped, for the most part safely west of the Tavronitis. A less fortunate section, perhaps forty in all, were carried out over the sea and drowned. The airfield was still being shelled by British gunners firing the old Italian 75; Les Young of the 21st watched the gunners in action:

> Our artillery must have been given a target because they started firing over open sights in the direction of the 'drome from where clouds of black smoke soon started to rise. At about 1600 hours a number of troop carriers crash-landed on the beach in positions not far from 21 Battalion. It was a clear case of the enemy being willing to sacrifice any number of planes to gain a foothold on the island. Lieutenant Rose, my platoon commander, and I, watched these planes crash-land and immediately afterwards burst into flames. I was never able to find out whether these planes were destroyed by the enemy or whether some mortars from another unit were doing some very accurate shooting.[9]

Student had rushed in a battalion of the 100 Mountain Rifle Regiment. If the boys from the Alpine pastures did not relish travelling by air then their reception fully justified these fears. The strip was stitched by fire and the lumbering transports suffered badly. Casualties were heavy, some planes preferred a rough landing on the beach but troops were being got in. The toehold had become, however precarious, a bridgehead. Despite their understandable wariness, the German report credits the air force for delivering Ringel's alpinists with considerable élan:

> The landing was carried out with the greatest dash and determination by Ju formations of Battle Squadron (Special Duties) 3, in spite of enemy artillery fire and to begin with, also machine-gun fire. A number of Jus were shot to pieces or burnt out on the beach and on the airfield. Extensive losses of mountain riflemen were avoided through the presence of mind of the pilots.[10]

For Student, still immured in his Athens hotel, there was little cause for increased optimism. The massacre of Nagele's men showed that the New Zealanders were well entrenched and in good numbers. He had to believe the Allies would now launch a counter stroke with every man they could muster and throw the invaders back into the sea. His own men of the Assault Regiment were utterly exhausted and those from the Mountain Division were still being shot up as they tried to land. Surely it was now only a question of when the blow would fall?

That night the Germans, watching anxiously from their positions, were treated to the sight and sounds of their amphibious reinforcement being shot up by the Royal Navy:

What we saw from [Great Castle Hill] was like a great fireworks display. Rockets and flares were shooting into the night sky, searchlights probed the darkness, and the red glow of a fire was spreading across the entire horizon. The muffled thunder of distant detonations lent sound to this dismal sight. For about twenty minutes we watched, until suddenly the fireworks ceased ... very depressed we returned to our Headquarters.[11]

If the attackers were downcast, the reaction at Creforce HQ was close to jubilation – this dazzling display of maritime pyrotechnics clearly heralded the demise of the seaborne threat the Navy had delivered. Freyberg turned to Brigadier Stewart and confided 'Well Jock, it has been a great responsibility'.[12] By this he appears to have indicated he thought the battle won. The obsession with the naval landings, which the misleading ULTRA intelligence had sparked, now seemed lifted. All that remained was to launch a vigorous counter-attack and recover Maleme. If, however, Freyberg believed the battle to be all but won, he was seriously mistaken. The fact that the Germans had been able to exert their grip on the airstrip at Maleme and begin, even shakily, to fly in reinforcements, meant that the crisis of the battle was only now at hand. If they could not be dislodged, then the night's action at sea and the sufferings of Cunningham's ships the following day, were purely anecdotal. The pivotal struggle was that for the control of the airfield. Quite simply, whoever held Maleme would hold Crete.

That Ramcke's advance had been stopped in its tracks and his eastern flank disposed of did not solve the essential problem; only a strong, vigorous and utterly determined counter-attack could recover Maleme and ensure the Axis defeat. With every hour that passed, regardless of loss, they would continue to grow stronger. Killing Germans, and in this the New Zealanders had excelled, just wasn't enough, the occupation had to be eliminated altogether.

By now the airfield had begun to resemble a scrapyard for wrecked aircraft. So many of those which had landed had crashed or been riddled as they unloaded. Amongst the battered survivors of the Assault Regiment there was no sense of victory and the newcomers from the Alpine Division found themselves hurled into a cauldron:

They [the aircraft] lay there immovable, like giant captured
birds, and slowly the work of destruction by the British Artillery
was completed. With the aid of a captured Bren Carrier a group
of paratroops and aircraft personnel did their utmost to clear the
Landing ground, but the wreckage still constituted a severe
obstacle to further landing Operations. If the following day were
to bring further losses on landing space the problem of supplying
the defenders would become acute.[13]

Brigadier Hargest had been entrusted by Puttick, as his immediate
superior, with the job of mounting the counter-attack. Both of these
were good soldiers, brave and dedicated but, as Freyberg ruefully
commented afterwards, they had been bred and trained for the
wrong war. Had Kippenberger been in command then undoubtedly
the day would have concluded differently but the 20th and 28th
Battalions, made available for the blow, were to be fought by an
officer whose ideas had been formed in the grinding attrition of the
trenches. There is also evidence to suggest Hargest was utterly
exhausted and, in less demanding times, would have been rotated
out of the line to recuperate.

The New Zealand line now rested primarily along a north-south
axis but with the northern flank extending eastwards parallel with
the coast road. The area was by no means secure, with pockets of
surviving parachutists barricaded into such impromptu strongpoints
as they had been able to occupy. Even if these remnants lacked an
offensive capability, they could still operate as a very effective thorn
in the flesh of the defenders.

Lieutenant Colonel Dittmer, commanding the 28th (Maoris),
suggested a preliminary sweep through the area to eliminate these
lodgements before moving up fresh troops for the attack, though it
was decided that it was now too late for such an initiative, eminently
sensible as the suggestion was. It was proposed that the blow would
be struck by the two reserve battalions alone, rather than launch a
sustained advance by the total of five which would then be in the
line.

By 1.00 a.m. the troops were to be at the start line, just to the west
of the Platanias River, with the assault on Pirgos timed for three
hours later. The 20th would pass north of the road and the 28th
south, their attack supported by a trio of the light tanks driving
along the road and forming the link. Having taken and consolidated
their grip on Pirgos the attackers would, after a halt of thirty minutes

push on to clear the airfield. Somewhat bizarrely, the orders provided for the 28th to fall back and allow the 20th to garrison Hill 107. Whether the other three battalions were expected to advance and consolidate is not clear.

That the operation be successfully completed at night went without saying: should the New Zealanders be exposed in open field once the sun rose and the prowling Luftwaffe held the skies, then they would be shot and bombed to extinction. In order for the 20th to move forward they had first to be relieved by the 2/7th Australians who, lacking trucks, had to wait until an adequate number could be assembled. This was complicated by the attentions of the Luftwaffe which unnerved many of the drivers.

Lieutenant Colonel Walker, commanding 2/7th, had entrusted his second in command, Major Marshall, with the unenviable chore of mustering an adequate number of trucks. As dusk was around 8.00 p.m., this would be the soonest it was deemed safe to begin the move west. The trucks should be ready some hours beforehand. The Major's task was a difficult one. Apart from the anxiety of the drivers, the Luftwaffe was also much in evidence:

> We turned a corner and found half a dozen planes above with the obvious intention of stopping us somewhere. I stopped the column until I was sure Savage with A Company had caught up and then we sailed on. It was rather exhilarating. The planes had now obviously got onto us, but the road was winding along a valley, and there were few straight stretches. The planes cruised about those stretches waiting for us ... twice I watched a plane single us out, bank and turn to machine gun us along the straight and I told the driver to crack it up. It then became a race to the curve ... We streaked along and I hoped the battalion was following.[14]

Having finally secured the transport and begun their move to the west past Souda, it was only too easy to become lost as they skirted the beleaguered port, riven by the incessant bombing. The trucks were obliged to play this deadly game with the swooping predators, when a mistimed burst of speed or an involuntary halt could invite an early death. As a result the leading elements did not come up with 20th Battalion till dusk, around 8.00 p.m. and it was much later before the final elements straggled in.

Freyberg, at 9.15 p.m., had given Puttick the strictest of injunctions that the 20th had to stay put until it was relieved. The General still feared a seaborne assault. Puttick was not about to disobey so stern an order. When more than two hours had elapsed and the threat had clearly been dealt with, these orders should have been seen as redundant. However, Puttick lacked either the confidence or the imagination to budge. Lieutenant Colonel Gentry, whose job it was to liaise with Hargest, was his GSO1:

> I was asked to go up to 5 Brigade Headquarters with one Major Peck commanding the light tanks with a view to giving any assistance that I could. At that stage there was little I could do and as I came back along the coast road to my great joy I saw the Royal Navy dealing with the seaborne invasion – a never to be forgotten sight – but there was no sign of 20 Battalion which I had expected to meet on the road. Puttick was asleep when I got back to Divisional HQ so I rang Inglis and told him about the battle at sea and asked why 20 Battalion was delayed. He told me that Puttick had given him orders that 20 Battalion was not to leave its positions until the relieving Australian battalion had arrived. As the reason for that order had now been overtaken by events I asked him to get the 20th on the road as soon as possible – even then it took far too long.[15]

By now Dittmer was fretting at the start line and it would take Major Burrows with the 20th a good couple of hours to slog forward to align for the attack. The cloak of darkness was infinitely precious and everyone on the ground was aware of the risks once its enveloping shroud was burnt off by the dawn.

Burrows had sought orders directly from Puttick to begin moving forward without attending on the relief but the request was denied. Dittmer had prevailed upon Hargest to repeat the same request but he also was refused, on the grounds, incredibly, that the position must be fully held whilst the possibility of a seaborne invasion remained. Puttick had, of course, seen ample proof that this threat had been dealt with.

Utterly frustrated, Burrows acted on his own initiative and ordered the companies forward as each was relieved. By 2.00 a.m. he arrived at Hargest's HQ; forty-five minutes later only two of his companies had come up. He felt it best to attack with what was available though this seemed to cause Hargest to 'wobble' about the viability of attacking at all. Puttick, when consulted, gave orders that the assault must begin – by now Dittmer's Maoris had been kicking their heels for over four hours. At 3.30 a.m. the tanks finally lurched forward, flanked by a section from the 28th to provide close support against grenades and Molotov cocktails.

No sooner had the advance commenced than the leading elements of both battalions ran into scattered German strongpoints, more numerous and better posted than had been anticipated, dug in behind solid masonry and with an abundance of automatic weapons. As Burrows recalled:

> ... [after] about half a mile we struck the first of the enemy posts. A German light machine gun, perhaps thirty yards away, opened fire with tracer bullets, followed immediately by tracer from other enemy posts across our front. The sudden shock of this outburst, the deadly hammering of several automatics firing together and the astonishing brightness of tracer bullets made men pause, even recoil, but not for more than a second or so. On my left and my right there was a sudden surge toward the enemy posts and not too long afterwards the shouting and the firing and the explosions of hand grenades died away and the advance began again.[16]

Private Melville Hill-Rennie found himself on the far left flank of the advance and he later related how:

> ... suddenly we ran into our first opposition. A Jerry machine-gun nest opened fire on us at a range of 50 yards and they got four of our boys before we could drop to the ground ... Jerry was using tracer and it was strange to lie there under the olive trees and see the bullets coming. I could see the explosive ones go off in a shower of flame and smoke as they hit the trees. We waited on the ground and finally the order came for my section to advance and wipe out the nest. We edged forward on our stomachs until we were within 20 yards of the Nazis who were tucked away behind the large tree and then opened fire with our one Tommy-gun, one Bren and eight rifles. As we kept up the fire the platoon officer cautiously crawled round to the side and slightly to the rear of the tree. Although it was still dark we could tell by the way the Jerries were shouting to each other that they didn't like the look of the situation. When we got round behind the tree the platoon officer jumped to his feet and hurled three Mills bombs, one right after another, into the nest and then

jumped forward with his revolver blazing. Single-handed he
wiped out seven Jerries with their Tommy-guns and another with
a machine gun.[17]

Most of these obstacles were encountered by Burrows' companies
from the 20th, north of the road, where the fighting was at close
quarters, desperate, confused and bloody. Dittmer and his Maoris
had a somewhat easier task – the surviving Germans had previously
been harassed by the guns of the 23rd and had withdrawn southwards, leaving only a few, but very troublesome, snipers. The net and
crucial effect was that by the time the sun rose in the east behind
them the men of the 28th were stuck at the crossroads north of
Dhaskaliana. Pirgos lay before them but the place was strongly held
and the 20th still lagged some way behind. The cloak of night had
vanished altogether.

Hill-Rennie remembered that as the 20th battled its way past
Pirgos, the blessed dark had given way to morning light. The combat
had been intense, the grinding, savage business of clearing one barricaded village house after another, each of the solid dwellings
converted with German thoroughness into a mini strongpoint. By the
time he and the other survivors had fought their way to within sight
of the airfield the Messerschmitts were already in the air and making
their presence felt. It was now 7.30 a.m. on one of those cloudless
and perfect Mediterranean spring mornings, when the bright, clear
light tilted the balance inexorably in favour of the invader.

Captain Dawson, acting as Brigade Major, retired to report to
Hargest on how matters now stood. He was not confident that an
attack could be launched in daylight with the vultures already
gathering in the skies above, like a horde of vast and deadly mos-
quitoes. No sooner had he gone, however, than Burrows with his two
companies came up. He and Dittmer decided to press on, C and D
companies of the 20th would attack the northern flank of the village
and press on toward the eastern edge of the airstrip. The Maoris
would assault Hill 107 while the tanks would storm the town head
on; a role for which they seemed unable to offer much enthusiasm.
For this they could scarcely be blamed, their thin armour and inadequate firepower would be pitted against 20-mm cannon and
captured Bofors deployed by the entrenched defenders. Almost
immediately the lead tank was hit and 'brewed up' – only the
wounded driver survived. Roy Farran, commanding the troop, had
dashed under cover amongst the bamboo to avoid being strafed and

had sustained damage to the tracking. It would be possible to repair his tank by cannibalising the one damaged – but this meant a long wait while the necessary mechanics could be found.

Without armoured support C company could gain no effective lodgement and D, while they reached the perimeter, took heavy casualties. The 23rd, detailed as a mobile reserve, did not lend any weight to the faltering attack and only one company followed the Maoris as they stormed Hill 107. Dittmer and his men performed bravely against an entrenched enemy, stronger in numbers, with a far greater weight of fire and closely supported from the air:

> We must get forward and get above and round the Germans whose bullets and mortar bombs were cracking round us. We could at times see German machine gunners running up through the trees. We collected in small groups and worked forward. Men were hit, men were maimed. The din of the fight was incessant. There seemed to be German machine guns behind all the trees. If we could silence one or two immediately in front we might break through.[18]

The fighting became increasingly confused and hand to hand. With banners, the arrogant blaze of the swastika, mounted on poles, the paratroopers counter-attacked in turn only to be driven back by these magnificent warriors, shouting their proud tribal challenges and sending the Germans scurrying back, '... the Huns with their fat behinds to us going for their lives down the gulley.'

Brigadier Hargest, still at his HQ, had come to the notion that things were going rather well and that the evidence of increasing numbers of German planes over and at Maleme suggested they were preparing to pull out. Even when Roy Farran and the remnants of his battered troop returned to confirm that Pirgos was likely too tough a nut, Hargest persisted in issuing optimistic signals. It must have been suffering from acute nervous exhaustion at this point and his reserves of endurance were utterly exhausted. The consequences of this lack of coordinated and inspired decision making at 5 Brigade were, ultimately, disastrous.

By midday the reality was that the attack had completely run out of steam and that the New Zealanders were depleted, exhausted and increasingly short of ammunition. Worse, they were vulnerable to a strong effort from the defenders, steadily reinforced by fresh companies from 11th Battalion, 85 Mountain Regiment, themselves elite troops.

By late afternoon Burrows, still with only three companies, had
decided to abandon the attack on Pirgos and to lend his remaining
strength to Dittmer's attack on Hill 107. The brave remnants of D
company could only watch, clinging to their toehold on the eastern
flank, as more and more German reinforcements were landed:

> The mortar and machine gun fire on the open ground was heavy,
> and we were lucky to get back alive. When we reached the
> 'drome the planes were landing (some leaving 'drome too) and
> the parachutists were jumping out and getting straight into battle
> for the Germans were counter-attacking on the right flank.[19]

There was little that the 20th could do to assist their hard pressed
comrades of the 28th. Despite their constant valour the ground
before them was swept with fire and without mortars no progress
could be made; casualties were heavy and mounting. After confer-
ring with Burrows, Dittmer decided he would return to Brigade HQ
and request Andrew and Leckie to come up in support. Both,
however, declined, stating that they needed to hold back and consolidate. Colonel Dittmer has not left a record of his thoughts at that
moment.

21st Battalion had begun their own advance in support at 7.00
a.m. and had achieved significant gains, moving westwards immediately south of Hill 107 and reaching the village of Vlakheronitsa, as
far forward as the positions D company of the 22nd had held on the
20th. They encountered relatively light opposition but, realising that
the advance further north had stalled, Colonel Allen halted to take
stock.

Realising the risk to their southern flank the Germans began to
strengthen the defenders in the village. Les Young recalled the
advance:

> Two companies of the 21st went forward without incident right
> into the village of Xamaduhori. It was apparent that any infil-
> tration that had occurred had withdrawn during the hours of
> darkness. We endeavoured to pass through the village but were
> met with a hail of machine gun fire and a fair amount of mortar
> fire. A number of casualties were suffered at this stage .... We
> received orders to withdraw to our previous position but the
> delay had been fatal – the enemy had appreciated the position
> and followed through smartly and it became a question of

running the gauntlet to get round and over the hills to our original positions. There were a number who did not make the grade.[20]

This was an opportunity lost. Hargest's conventional plan of attack had laid the emphasis on the heavily defended coast route whilst the enemy's flank hung in the air. Had the thrust been offered as a feint and the weight of the attack developed around and to the south, then this might have proved decisive. The history of the campaign is, of course, littered with so many 'what ifs'. The plain fact was that the counter-attack had completely stalled; when Andrew and Leckie refused to support Burrows and Dittmer the chance was lost. It was a dire blow and must have been doubly so for the survivors of the two forward battalions; that Crete was lost was no fault of theirs.

Beneath the strengthening sun, the harsh light mocking the hopes of the night, the advance elements of the 20th had battered their way forward to the eastern rim of Maleme aerodrome. It was a case of so near and yet so far. Their loss had been grievous and the mocking orb drew in the omnipresent swarm of Me109s. There was little more that could be done. Lieutenant Maxwell, the single officer still on his feet, remembered with bitter gall:

> We reached the clear part of the 'drome all right – there was stacks of aircraft, some crashed, some not – I remember P. Amos saying 'I've carried this anti-tank rifle all the way and I am going to have one shot'. He fired two shots into one aircraft and made a mess of it. Broad daylight – at this time we had come under most intense mortar and machine-gun fire with the clear ground of the 'drome in front of us. I pulled the Company back about 100 yards, back into the cover of some bamboos.[21]

Charles Upham, who had performed so valiantly, shared the bitterness and frustration felt by the men on the ground; those who had fought so hard and achieved so much despite bad planning and worse timing:

> If only the attack had gone in at the right time we'd have had the hours of darkness when we were very successful, and it was only when it got to eight o'clock in the morning that the Germans got on top of us. We were completely successful until then, we advanced right through the aerodrome, some fellows went right past the aerodrome, along the beach, we got right up by the

aerodrome where the Germans were in force. But once it got daylight these fellows had a lot of machine guns and there was a lot of them, whereas the ones we were meeting in the dark were sort of panicky.[22]

Sporadic fighting continued all day but, as the fatal sun began to wane, it was obvious the German grip on Maleme was both unbroken and, by now, unbreakable. The Allies had been harassed by the Luftwaffe throughout the long day and, by the time the shadows began to lengthen, the invaders felt sufficiently emboldened to throw out a cordon of fighting patrols.

Swinging onto the offensive for the first time, the alpine troops, well supported by artillery, struck at the Maoris of the 28th, on the left of the battered remnants of the 20th. Despite the battering and exhaustion the New Zealanders rose from their meagre cover as the mountain troops crested the ridge. Led by Major Dyer who was armed with nothing more than a walking stick, they hurled themselves forward. Dyer himself left a most vivid description of the fight:

A scattered band of dark figures under the trees ... with knees bent, and leaning to the right they slowly advanced firing at the hip. They did not haka, for this was not rehearsed. Instead, there rose from their throats a deep shout 'Ah! Ah! Ah!' as they advanced. Then the cartridges in their magazines being exhausted, they broke into a run with bayonets levelled and their shouts rising as they went ... And the pride of the German army turned and fled.[23]

Four years later, in the course of his trial, Student had this to say: 'When on the 21st May all the reserves had jumped and conquered the aerodrome of Maleme, from that time the battle for Crete was won for Germany.'[24] In this assumption he was to be proved entirely correct. Despite the disastrous events of 20 May, Student had perceived immediately the fact that the bare foothold gained at Maleme on that crucial first day held the key to eventual success. His keenness and the inestimable advantage of superior communications ,coupled with the exemplary manner his subordinates carried out his orders, indicates a strength and unity of command, a level of strategic understanding far in excess of that possessed by the Allies.

Student's calm appraisal is made with the benefit of hindsight and, though correct, his appreciation was not immediately clear to the German troops at Maleme, who still considered their position extremely precarious. Fresh detachments from 5th Mountain Division were being flown in to consolidate the expanding toehold and the indefatigable Major Wenning was, once again, in the air:

> As the high coastal mountains of Crete appear over the horizon and we can slowly recognise the markings of the bay of Canea we can see in the middle of this a huge, black tower of smoke. That must be the airport at Maleme ... covered in vast smoke and dust clouds, a large number of Ju52s circling above it. Coming closer we realise the reason for this circling. On the airfield which is extremely small there are lying a number of crashed planes and some of these are burning. One Ju52 which is landing collides with one of these wrecks, spins around and lies motionless in the middle of the narrow, inadequate landing strip. This diminishes our chances of landing especially as I can count ten other planes in the air which were circling before we arrived. Despite this my pilot tries to land three times without success because during each approach to the strip suddenly another plane turns in from the side, cutting us off. On the airfield itself we can see shells exploding. The enemy artillery is pouring an accurate fire onto the runway, their batteries somewhere in the mountains and well disguised because we can't make out the muzzle flashes although we try hard. For this reason already this morning several Ju52s suffered direct hits while landing causing many dead and wounded. All in all a rotten situation.[25]

On the ground, the equally formidable Lofty Fellows was also in action:

> I had a Bren gun and two loaders, a good position, it was a dugout, roofed in with a fire opening ... and these two chaps were fantastic, they had a pair of binoculars and being artillery lads they could pick up movement, mortar crews being put into position and all this sort of thing and they were only too pleased to tell me where to fire, not that I could see anything, and they'd say, OK it's 1500 yards or right of the bridge entrance or so many degrees and I'd bang a few shots away, and there was a Vickers gun up the hill somewhere and when they saw our stuff hit the ground they'd start up theirs, and there was a Naval chap on top of the hill and he'd supply the odd 4-inch shell just to stir things up. So we had quite a good time. The Germans put an attack in to try and shake us out but with all our firepower they

decided to retire and they went back up the beach so I wound up the sights of the Bren gun and just let in rip just to help them on their way. It was very pleasing to watch them ducking and diving. I believed in firing guns as long as they'd fire and I kept on all afternoon with a four gallon can of water alongside. They always reckoned you could throw the barrel of a Bren gun into water to cool it, pull it out and start again, and they were right, you could. I don't know how many rounds I fired but I was continuously reloading it and firing even after it got dark.[26]

For the crack troops of Ringel's division their arrival on Crete was anything but tranquil. The planes barely touched down and the soldiers bundled out into the baking dust, a dread chorus of mortars and small-arms fire, their raucous greeting. Victory seemed far from being assured. Lofty Fellows witnessed one of the RAF's infrequent sorties as a pair of Blenheims strafed the overcrowded beach. Wenning's plane had ditched quite close to the waterline. The major, though temporarily knocked unconscious, emerged unscathed as did his shaken passengers. Wenning made his way along the shore, hitherto a tranquil backwater but now littered with the detritus of war. He made sure to wear his sunglasses and adopt a suitably nonchalant air. He found that prisoners had been put to work to level out the moonscape of craters. Apparently, when some had demurred, Major Snowatzki had provided motivation by shooting three out of hand.

The first serious Allied attempt to counter-attack was mismanaged at Brigade level and, despite the extraordinary valour of the main two attacking battalions, it had failed. A prime opportunity to wrest back the initiative and win the battle had been squandered. The Germans, regardless of the risk to men and machines, did not intend to squander the respite. Nonetheless, as the passages quoted above show, the invaders at Maleme were far from having things all their own way. British artillery, well dug in and despite the indifferent quality of the captured Italian 75s, continued to harass and delay the build-up. Lofty Fellows and his fellow machine gunners guaranteed the Germans a warm welcome.

As the spirited attempt by 20th and 28th Battalions ran out of steam in the afternoon of 22 May at 5.00 p.m., Freyberg called a fresh conference and threw his full weight behind orders for another assault. This time the attacking force would comprise the whole of 5 Brigade, strengthened by the addition of the 18th New Zealand and

2/8th Australians. Puttick appeared ready for the challenge but suffered a fresh crisis of confidence when, on reaching 4 Brigade HQ, he received intimation of renewed German aggression in the Galatos sector.

The euphoria which had earlier buoyed Hargest had been replaced by a profound pessimism – he did not consider 5 Brigade could do any more. The mood was infectious and Puttick was sufficiently alarmed to order the complete withdrawal of the entire brigade. Freyberg was most reluctant to accept the need for further retreat as this would effectively neutralise any prospects of renewing the offensive.

He had become increasingly worried as the afternoon of the 22nd wore on. Wavell had refused further reinforcements and doubts were beginning to crowd the General's tired mind. He had commanded the New Zealand division all the way through the heartbreak of the Greek fiasco and its continued survival was now at stake on Crete. He was being pulled in two directions. He had his duty to Wavell, to Churchill and the Empire and, at the same time his obligation to the home government. New Zealand did not have unlimited resources of manpower on which to draw, the division represented the cream of the Dominion's manhood.

It was a heavy burden for any man to bear and the General had had doubts about the viability of the defence from the outset. The Germans dominated the skies, making daylight operations virtually impossible. He had men but these were deployed along the long strip of the western flank of the island. He lacked transport, armour and artillery support. With the enemy now tightening his grip on Maleme the pendulum was inexorably swinging in his favour.

The news from his forward divisional and brigade commanders, Hargest and Puttick, was dispiriting, they told him the men were exhausted and demoralised, that the link between 4 and 5 Brigades had been severed and that the Germans were probing, in strength, toward Galatas. In fact most of this was untrue. The New Zealanders were full of fight and not the least bit despondent. The surviving parachutists in Prison Valley were not an immediate threat, effectively neutralised. In tactical terms the time was still ripe for a determined assault along the line 21st Battalion had taken, to hook around to the south of Hill 107 and pinch out the salient around Maleme.

Neither officer had gone forward to view the situation first hand and no coherent plan for a further counter-attack had been prepared.

After the war Kippenberger was not alone in criticising his superiors, citing their pessimism, lack of offensive spirit and lack of any real desire to seize the initiative. 'Kip' concluded that these were serious strategic limitations which had a profoundly damaging effect. It is very difficult to disagree with these sentiments, even making due allowances for the physical exhaustion of both men and their conditioning from a previous war. They did not, to their credit, wish to expose their men to the needless slaughter witnessed so often in the trenches, but it was their ill-founded negative sentiments that were now threatening to unravel the campaign.

Consequently, Puttick spoke to Freyberg by telephone and advised he felt that it was simply not possible to mount further attacks and that it would be more prudent to withdraw. Puttick, quite rightly, did not want to punch a salient into a strengthening enemy position, simply to see it crushed by superior enemy firepower. Freyberg could, at this point, have insisted, however, he did not. He was always disinclined to overrule his subordinates and whilst it is invidious to criticise a man for being too decent, wars are not won by decency. Napoleon Bonaparte was not a nice man, but he would not have hesitated to demand that the attacks go in.

Perhaps the real reason lay in Freyberg's own doubts. He was, as we have seen, prone to severe mood swings and the early euphoria of 20 May, the destruction of the German armada, had worn off to be replaced by the increasing certainty of defeat. It is impossible to imagine just how sapping was the constant fact of German air superiority. The Luftwaffe sorties were relentless and effective, with movement restricted to the short hours of darkness. Despite the large numbers of Allied personnel on Crete, many of these, in terms of their combat effectiveness, were bouches inutiles.

Freyberg thus concurred, sending Brigadier Stewart forward to liaise with Puttick so that a detailed plan of action could be prepared. Every minute that now passed strengthened the Germans and weakened the Allies. By this decision the loss of Maleme and thereby of Crete was assured, whatever happened thereafter became purely anecdotal. What remained to be decided was simply the magnitude of the disaster.

When the orders were hammered out just after noon, the whole of the 5 Brigade was to withdraw to the line held by the Maoris just in front of Platanias. This implied that the Allied guns could no longer fire upon the aerodrome at Maleme, the German foothold thus became their vital bridgehead. Dittmer greeted this news with horrified astonishment, the men did not expect any such order, they saw no reason for withdrawal. Despite their rebuff the day before, they considered that matters, overall, were still going well. The decision to retreat was not occasioned by any failure in the forward units, quite the reverse, it was made as a result of command failures, well behind the front line.

They had, thus far, met the detested enemy, who had hounded them from Greece, still on unequal terms as his planes ruled the skies and swooped at will, but had shown they were more than a match for him in an infantry battle. They had taken on and, in some cases, slaughtered, the pride of the Axis. In spite of all this the stink of defeat began to raise an ugly miasma.

It was not until the following dawn that Colonel Dittmer received his orders, and he immediately contacted the 23rd to seek confirmation. We can only imagine his chagrin as he had to pull his proud battalion from the ground they'd fought so hard to win. On the flanks of the 28th the withdrawal was conducted under enemy fire. Lofty Fellows, inevitably, was in the thick of it:

> I'd acquired a German first aid pack and I'd thought well, I'll hang on to that and give it to the doctor, and also a flask of rum. It was quite a warming thing to have, a sip of rum every now and then. On the way up the hill was a Sergeant, Charlie Flashoff, who'd been wounded in the legs with a grenade and I said; 'Would you care for a little libation' and he said he would so I sat down and unscrewed the cap, filled up, went forward to give him a drink and there was a clatter of firing practically down my neck and I felt something thump into this medical satchel thing I had. I decided that wasn't the place for me, and poor old Charlie was on the ground anyway. So I took off down the hill and the only thing that stopped me in flight was two chaps in a slit trench with a tin of peaches and I was loath to pass them, food being one of the staple things I needed, so I joined them. And one said, what the devil's that running down the back of your pack, so I took it off and I found four bullet holes in it.[27]

Quite a close shave, and Lofty was right to stop and share the peaches; adequate supply of rations was already becoming problematic. This difficulty of supply was occasioned primarily by lack of transport, a poorly organized commissariat, exacerbated by the incessant prowling of the Luftwaffe. Food would become an increasing preoccupation of hungry men as the battle persisted.

In some cases the precious 75s, which had rendered such sterling service had to be abandoned as had a couple of 3.7-in. howitzers. There was nothing more dispiriting for a gunner than to have to destroy his own gun and find himself suddenly reduced to the status of a mere infantryman! The Germans were pressing hard now, sensing their advantage, dragging up the captured Bofors, and Dittmer's Maoris were involved in at least three skirmishes. Those too badly wounded to be easily transported had to be left; having fought so hard, it was galling for the brave New Zealanders to see their comrades left behind.

Major Thomason of the 23rd had also received orders to pull back:

> .... Received orders to withdraw to the Platanias River area. I led the Battalion back through a route I had previously checked if we had to withdraw. We had to leave our wounded behind as we were unable to get any transport through. A German officer prisoner undertook to contact his side and advise them of the position of the FAP and the wounded to save them from unnecessary attack. Both our doctor and padre stayed behind with them, a rather unfortunate decision as regards the doctor as we had great need of him later.[28]

Les Young, with the 21st, had a similar unpleasant experience; the Battalion found itself:

> ... in what appeared to be a most unenviable situation. The Hun seemed to be getting into a position on three sides of us. This fact was probably appreciated by others as we received the order early in the morning to withdraw due east until we crossed the first river. This was not far, being a distance of approximately three miles, but for weary troops scrambling over hills carrying what little rations and equipment they had proved quite an effort. We eventually arrived at the west bank of the river and found it to be nearly waist deep. Troops plunged in without ceremony and proceeded to cross. An enemy machine gun post composed of Huns who had infiltrated from where they had landed in the prison area had established themselves on a high rocky pinnacle just south of where we were crossing the river and caused considerable trouble and casualties to our withdrawing troops. Extreme difficulty was experienced in getting wounded back and many were left at FAPs further west and were eventu-

ally taken prisoner. We established ourselves on the east bank of the river about 1,000 yards west of Platanias in a line commanding two miles inland and straggling south along the river bank.[29]

Part of the problem with the hasty decision to withdraw lay in poor communications. The lack of serviceable tactical radio reduced the giving of orders to Great War levels:

> Here was a case where we suffered through lack of modern equipment. A battalion fully equipped should carry portable wireless sets. We had none. The Germans had swags of them.[30]

Herein lay the rub. Not only did the invaders possess more and better automatic weapons, but their chief advantage lay in their abundance of tactical radio – they could communicate. This gave them an inestimable advantage, particularly in the confused nature of the fighting on Crete.

Major Jim Burrows did not get his orders until 4.30 a.m. – these were instructions that he should have received over six hours earlier. By the time his unit was on the move it was daylight and so he wisely chose a more circuitous but less exposed route back to Platanias. Here they did not halt but, after re-grouping, pressed on toward Galatas:

> We moved to Platanias in small groups. Nothing much happened to my group and we arrived without casualties. Some groups were caught by machine-gun fire and some by planes. We had to cross a deepish river and our group got safely across a swing bridge. Jack Bain and his men however had to wade as a machine gun had ranged onto the bridge. They came in wet to the neck.[31]

Group Ramcke were snapping at the heels of the New Zealanders, who never failed to give a good account of themselves in the savage little rearguard actions which raged like sudden bush fires across the arid ground. By noon on the 23rd, according to their own sources, Ramcke's leading units were in contact with the defenders holding the new line at Platanias. Their report is coloured by accusations of guerilla activity, 'civilian snipers' who were accused of mutilation. The Cretans were certainly unbending in their resistance to the hated invader though talk of mutilation is almost certainly propaganda.

Captain, later Colonel Tsatsadakis, a native of the village of

Episkopi, which nestles in the hills above Heraklion recalled the local response:

> We knew about the war in mainland Greece from newspaper and radio reports so we were expecting some sort of German invasion by air or by sea although we didn't know when or how. When we saw the paratroopers start to come down all the men in the village got their guns. Many we killed around the church where they came down and after we buried the bodies in holes in the ground. One though I saved and didn't kill and after the capture of the island he came to our village and stopped other Germans from burning down the village with petroleum as a reprisal. And too we reburied the German dead in proper graves before the other Germans came, on the instructions of our priest who was a clever man, and put little bunches of flowers on the graves, so the Germans wouldn't harm us for killing their soldiers.[32]

The village priest was a clever man indeed. His quick thinking undoubtedly saved his parishioners from the full horror of Teutonic vengeance which was to cost many of these brave Cretans dear. By mid afternoon on the 23rd Ringel was able to signal Student, still relegated to a spectator's view back in Athens, that his division had established contact with Heidrich's survivors in Prison Valley and, even more importantly, had completely secured Maleme airfield.

German morale was rising, even the slowest witted could sense that the tide was turning. The converse was, naturally enough, true of the Allies. Lofty Fellows, still in the thick, became aware of this, as he moved eastwards from Platanias. At one point he was:

> ...sitting among the olive-groves with my Bren and feeling very dirty. I'd lost all my gear and couldn't have a shave. After feeding myself, all the time I was a soldier I tried to have a wash and a shave every day. It does wonders for morale. I looked over a bit further and about thirty yards away under the olive-trees there was an officer and he was having a shave. So I scouted round the back and came up to him and I said: 'How about letting me have a go with that thing when you've finished', and he agreed. I was sitting there shaving away when suddenly somebody shouted: 'Get down can't you, stop moving around, you'll draw the enemy fire.' Gave me such a fright I nearly slit my own throat.

I found out later he'd been doing it all the previous day and that he was the captain in charge of the company. It was terrible for morale. All these men looked to him for guidance and all he did was give them the nervous jitters about German aircraft. And he wasn't the only one. In some places if you moved while the aircraft were strafing somebody on your own side would take a potshot at you. Stupid really because the planes were coming in so low that we could have potted them easily with a Bren or a rifle. If we'd shot back instead of lying low we might have brought down a lot of planes. But instead we had these fools shouting: 'Don't move, lie still, don't attract them!' and all the rest of it.[33]

As an immediate consequence of this easterly withdrawal the Maleme front now merged into the overall situation around Prison Valley and Galatas. Here little fighting had taken place since the clashes on the first day. The German survivors had been left unmolested but with only meagre supplies reaching them. Heidrich commanded the remnants of his three battalions from 3 Parachute Regiment plus the Engineer Battalion. Ammunition, food and fresh water were in very short supply.

To try and establish contact with Ramcke he had formed an ad hoc brigade from survivors of his 3rd Battalion and the engineers who were deployed over the lesser reaches of Signals Hill, effectively in no man's land. It was likely that it was this manoeuvre which had disconcerted Puttick on the afternoon of the 22nd.

At the same time Heidrich put in two attacks aimed at securing a lodgement on the Galatas hills though the attackers were not numerous; a few companies in each assault group. They were well supported by mortars and strafing Me109s. The orders to attack were received without much enthusiasm by the tired and hungry paratroops. Their numbers, it seemed, were too few, memories of their earlier losses still fresh.

When, however, Major Derpa, commanding 2nd Battalion, questioned Heidrich's orders, the Regimental Commander, exhausted and stressed, rounded furiously on his subordinate going so far as to accuse him of faintheartedness. Von der Heydte, a witness to this unseemly exchange, recorded how the Major, still rigidly to attention paled, taut with fury, but replied with dignity that whilst careless of his own life he had a responsibility to those he commanded. Derpa fell, mortally wounded, in the course of the assault.

As the attack closed around the flank of the gallant but hugely outgunned Petrol Company, the Axis gained the summit of Pink Hill where they quickly established machine-gun nests. At this point the situation seemed critical, the battalion reserve was no more than platoon strength when, quite suddenly, the day was saved by the intervention of none other than Captain Forrester of the Buffs, late of the Military Mission, leading a scratch force of Greek soldiers and Cretan auxiliaries:

> There came a terrific clamour from behind. Out of the trees came Captain Forrester of the Buffs, clad in shorts, a long yellow army jersey reaching down almost to the bottom of the shorts, brass polished and gleaming, web belt in place and waving his revolver in his right hand. He was tall, thin faced, with no tin hat – the very opposite of a soldier hero; as if he had just stepped on to the parade ground. he looked like a Wodehouse character. It was a most inspiring sight. Forrester was at the head of a crowd of dis-orderly Greeks, including women; one Greek had a shotgun with a serrated edge bread knife tied on like a bayonet, others had ancient weapons – all sorts. Without hesitation this uncouth group with Forrester right out in front, went over the top of the parapet and headlong at the crest of the hill.[34]

The wild charge of the immaculate Forrester and his ragged heroes was an incident straight from the pages of Dumas or G.A. Henty; the kind of inspired derring do that Churchill rightly loved. In the immediate tactical context it stemmed the tide. A second German thrust, to the east around the vital bastion of Cemetery Hill, was seen off in a similar stylish fashion by more Greeks led by Captain H.M. Smith.

By 2.00 p.m. on the afternoon of the 23rd the retreat had been completed and the German forces advancing from the west could link up with Heidrich's detached contingent.[35] As the day wore on the German fire increased. This was bad enough but both brigade and division were more fearful of a general outflanking move from the south – the line before Platanias was already looking untenable. For Hargest and Puttick this must have seemed a replay of the Greek nightmare; their troops, massively outgunned, facing an enemy who was growing steadily in numbers with full tactical air support.

That night, with the fear of encirclement growing, the weary New Zealanders would be obliged to consolidate on the Galatas heights, '...behind the 4th and Kipp's 10th Brigade'. The battle could not now be won but neither was it yet lost. The Germans had sustained fearful losses but there still remained a possibility that some form of deadlock might ensue, until the inexorable demands of Operation Barbarossa led to a withdrawal.

Churchill had already cabled Freyberg that: 'The whole world is watching your splendid battle on which great events turn.'[36] He had also urged Wavell to continue the battle as a means of tying down enemy forces, thus enabling him to consolidate the situation in North Africa. More improbably, he then suggested sending more reinforcements including tanks, in the belief the Axis were at their last gasp. On the ground the fight for Maleme was over. The battle of Galatas was about to begin in earnest.

> It [the position] was in the trees and was potholed with short trenches. All around were unburied bodies of Cretan soldiers. They could not be buried because there appeared to be no digging equipment. It was because of this lack that we could never enlarge any trench system that we occupied. We were very close to the Germans at this stage and movement was very restricted during the daytime. However, the enemy kept their distance apparently preparing for the final attack on Galatas. The build-up came by the ranging of the German mortars. First they dropped a cluster of mortar bombs on one side of the little valley up which the road ran; then a cluster up the other. The cluster on one side was right on our position and the day was a long, long one. I remember Lieutenant Dill saying: 'Tomorrow we are for it. I'm going to find myself a bath' and he went away into an empty house nearby and had his bath.[37]

As the New Zealanders withdrew first to the Platanias line then onto the heights before Galatas, the Germans increased the build-up at Maleme, wasting no time in repairing the ravaged airstrip, now finally free of the attentions of the British gunners. Before the day was out the first Me109s had arrived as had two batteries of 95 Mountain Artillery Regiment, 95th Anti-Tank Battalion bringing a score of 50-mm guns, the first infantry formations from 141 Mountain Regiment and 55th Motor Cycle Battalion.

Soon the entire divisional strength would be deployed and the balance not just of numbers but of firepower would swing massively in favour of the Axis. The battle was now directed personally by Ringel, who had flown in late the evening before. The fight for Crete was no longer about vertical envelopment, it was to be a straight contest between conventional forces, even if it was the airborne arm who, at such catastrophic cost, had prised the door open and held it wide for just long enough.

Student's participation, humiliatingly, was reduced to the role of spectator. He would have to suffer the gall of watching Ringel, the plodder, consolidate what should have been his victory. The Austrian informed the surviving paratroops that all forces remaining in the west and centre were under his direct command – Ringel Group. Ramcke had the job of consolidating the survivors from Maleme into a single battalion which was detailed to secure the left, coastal flank of the advance. This was now a Wehrmacht, not a Luftwaffe operation. Whatever was salvaged would be to Ringel's credit and not that of Student whose star, which had once flashed so brightly in the higher echelons of the Nazi command, had now dipped irrevocably and would soon vanish entirely from sight.

Ringel identified four main priorities:

1. To secure Maleme airfield
2. To occupy Souda Bay
3. To rescue the survivors at Rethymnon and Heraklion
4. To secure the occupation of Crete and the final defeat of the Allied forces.

Ironically, it was the rigid adherence to these orders that would enable the final withdrawal to and evacuation from the south coast by Sphakia to succeed. Ringel was determined to secure the island from west to east rather than north to south. Galatas was to be the target of the General's next move, a strong and concerted attack from three sides, the Alpine troops bearing the brunt in the centre, supported by a powerful thrust on the right flank, using two fresh battalions to snuff out the Greeks holding Alikianou and then to cut the coast road in the rear of the New Zealanders.

On the Allied side Freyberg had caught some of Hargest's angst. He felt that the continuous harassment from the air and the gathering weight of German ground attacks would prove too much for his tired men. It may well have been at this point that his mind turned to what, a couple of days previously, would have seemed unthinkable – the need to consider how best to save whatever forces could be saved. London, increasingly outpaced by events, was urging Wavell and Cunningham to make further efforts: '... great risks must be taken to ensure our success.'[38]

If a stand was to be made then the range of low hills around

Galatas was an excellent choice as the topography favoured the defender; the sweeping bracket of hills before the town begins in the north where Red Hill rises beyond the ribbon of the coast road. The ridge that bends east encompassed Wheat Hill, Pink Hill and, at the eastern extremity, the lump of Cemetery Hill that pushed forward into the dead zone of no man's land. The arc of this high ground was bisected by Ruin Ridge, with the crest of Ruin Hill rising to the west.

On paper Puttick's forces were more than adequate but he had withdrawn most of the fighting formations to the rear – the whole of 5 Brigade. No use of potential reinforcement from Chania/Souda had been proposed, so the actual defenders comprised the various ad hoc formations which had, with considerable gallantry, secured the line till now. These were, on Pink Hill, the remnant of the Petrol Company, with Major Russell's improvised detachment of largely support personnel. Further north lay the Composite Battalion holding the ground from Wheat Hill to the shore.

It may be surmised that Puttick intended the fight for Galatas to be no more than a holding action while a further withdrawal was in hand. He had positioned two fresh, first class battalions, both Australian, 2/7th and 2/8th, as the 'hinge' on which such a further retreat would turn.

The weary defenders could clearly discern the steady reinforcement of Heidrich's survivors as the 23rd wore on. The paratroops would not, in fact, be expected to play a significant role in the attack. They were regarded by their replacements from the Mountain Division as being played out, not that the Alpinists had any sympathy for the cream of the Luftwaffe.

If the New Zealanders possessed one vital asset it was Howard Kippenberger. It was he who, during the night of 23/24 May, persuaded Puttick to release 18th Battalion to shore up the right flank and to place the defence under Inglis' command, making the battle very much a 4 Brigade affair. With the Brigadier keeping his HQ back toward the coast road and Kippenberger ranging along the line, there was little more that could be done.

With the Composite Battalion being withdrawn to the line of Ruin Ridge, Gray's 18th took up positions on Red and Wheat Hills but left out Ruin Hill. Grey simply did not have sufficient troops to hold his extended line, which reached from the coast in the north-west to the ground held by Russell Force to the east. The opportunity was not wasted and, during the 24th, three German companies from the 1st Battalion of 100 Mountain Regiment dug in on the reverse slopes and positioned their heavy mortars to enfilade Red Hill. If they could clear this then they could repeat the performance and drive the defenders from Wheat Hill in turn.

Throughout the day the 18th fought with courage and resolution, closing on the flanks of the new salient. So fierce were they that Colonel Utz, bearing in mind Ringel's maxim 'sweat saves blood', decided to leave Galatas well alone until the Stukas could be called in the following day. Despite the stout resistance offered by the New Zealanders, their numbers were still few.

When Kippenberger walked the line of the Composite Battalion's new positions next morning he was keenly aware that both officers and men were exhausted. Moving reserves up from east of Galatas during daylight hours was virtually impossible, all that was immediately available were the tired remnants of 20th Battalion. Even the aggressive Kippenberger felt an intimation of impending disaster.

In contrast to their depleted adversaries, the Germans, since Utz had drawn a halt on the 24th, were building up their strength so as to be able to launch a concerted series of attacks all along the line; exactly the level of pressure Kippenberger feared. In echelon from the north, Ramcke Group would attack with two battalions, then 2nd Battalion of 100 Mountain Regiment, the 1st in close support and finally, the remains of Heidrich's paratroopers on the southern flank. Utz, leading the Alpine troops, would deliver the main blow, his battalions augmented by the addition of all of the anti-tank and artillery formations – a most powerful spearhead.

Against this formidable firepower Colonel Gray had only a handful of mortar bombs remaining. Upon request for re-supply he was issued with thirty more – all there was to be had.[39] The Germans had conceived a high regard for these stubborn and resilient New Zealanders and were cautious about throwing men into the attack without air support – 'sweat saves blood'. Ramcke's brigade had called down Stukas in the morning and Utz spent a frustrating afternoon awaiting his. As there was a marked and understandable reluctance to force the issue on the ground with the risk that the attackers were then caught by their own bombs, the afternoon wore on with only sporadic fighting.

The twin bastions of Wheat Hill and Pink Hill were crucial objectives of a successful assault and dominated any attempt to infiltrate over the lower ground. In the late afternoon, when the Cretan sun was at its strongest, the pressure of Ramcke's attack in the north, sustained by overwhelming fire support, began to wear down the defenders in that sector. Inglis sent in his two reserve companies, all the 20th had left to offer, to seal the breach. Even the battle hardened *Fallschirmjäger* were awed by the ferocity of these New Zealanders who simply refused to give up their positions regardless of the weight of bombardment.

With his final reserves now battling in the line, Kippenberger faced the full weight of Utz's attack once the Stukas had again assaulted Galatas; the siren wail of these deadly dive-bombers, competing with the shock of exploding bombs and the incessant crack of the Axis mortars firing from Ruin Hill. Into this inferno Utz committed his elite attack formations. As he was receiving repeated requests from the harassed defenders of Wheat Hill for permission to retreat, Kippenberger went forward to see for himself the effect of the German bombardment:

> [I] went forward a few hundred yards to get a view of Wheat Hill and for a few moments watched, fascinated, the rain of mortar bursts. In a hollow, nearly covered by undergrowth, I came on a party of women and children, huddled together like little birds. They looked at me silently with black, terrified eyes.[40]

The pressure was intense, by sheer weight of metal and numbers Utz's battalions were able to bludgeon their way forward. Every man who fell amongst the defenders was irreplaceable.

Captain Bassett, attempting to move within the New Zealanders' shrinking perimeter, found the effort:

> Like a nightmare race, dodging falling branches, and I made for the right company and got myself on their ridge only to find myself in a hive of grey-green figures so beat a hasty retreat sideways until I reached Gray's headquarters ... He greeted me with 'Thank God Bassett, my right flank's gone, can you give us a vigorous counter-attack at once?' ... a nest of snipers penetrated into the houses pelted at me and a Stuka keeping a baleful eye on me only (or it so it seemed) cratered the road as I scuttled.[41]

By early evening, around 6.30 p.m., Gray himself went forward to try and steady the line but with both flanks ready to fold. He managed, by this valiant example, to stem the rot but, all too soon, the dazed and exhausted men were reeling back in a steady trickle which threatened to turn into a spate. Within half an hour the defenders' positions on Wheat Hill had been forced and the companies flanking the position driven in. A major breach now gaped.

Then came Kippenberger:

> Suddenly the trickle of stragglers turned into a stream, many of them on the verge of panic. I walked among them shouting 'Stand for New Zealand' and everything else I could think of. The RSM of the 18th, Andrews came up and asked how he could help. With him and Johnny Sullivan, the intelligence officer of the 20th, we quickly got them organised under the nearest officers and NCOs, in most cases the men responding with alacrity.[42]

Such was the heroism of a single man at the moment of crisis that others would never forget the image or the inspiration, '...in the middle of it all this mighty, mighty man'.[43] The Brigadier's intervention helped to stem what was becoming a rout and the men fell back in as orderly a manner as the weight of German fire would allow. As the relentless pressure seemed only to intensify, the southern companies bent and then folded back toward Galatas itself.

> German infantry advancing under a heavy barrage of mortar and aerial strafing, completely blanketing our positions, and when the barrage lifted they were right amongst us with machine and Tommy guns, putting up a murderous fire. We got the order to retire and how we got through such a hail of lead I don't know yet. As we made our way back in small groups the planes got in among us strafing, the bullets clipping the bark and branches from the trees all around us. Our line of retreat to Galatas was being bombed so the obvious thing to do was to bypass the village. Two friends and I stuck together and got back all right to the next line of defence.[44]

Lieutenant Dill, he who had earlier sought a bath in the foreknowledge of the weight of the blow that was about to fall, was amongst those mortally wounded, and was dragged back to join the swelling horror around the Regimental Aid Post, now deluged with scores of wounded men. Kippenberger was forming a new defensive line along the crest of the Daratsos Ridge which angled north-eastwards before the village of Karatsos. Colonel Gray, 'looking twenty years older than three hours before', was among the last to retire.

A frantic message to Inglis, borne by the redoubtable Captain

Dawson, had produced a motley assortment of last ditch reserves, the 4 Brigade Band, a concert party, a rather random, divisional music outfit 'half orchestra – half NAAFI' and a pioneer battalion. The only combat ready formation to appear was a single company from the 23rd – immediately deployed to forge a hinge with the remnants of the 20th clinging to the right flank.

In the midst of the shooting the gallant Sergeant Andrews was hit in the stomach. Kippenberger records, with regret, how he felt he should never see this stalwart again. In fact, Andrews survived and the pair met up again in Italy in 1944.[45] For a brief moment, in the true course of this whirlwind battle, a sort of calm appeared to descend and hopes were raised that the Germans had, at least for this day, shot their bolt; far from it.

The truncated line was a rough wedge shape; on the long southern side, the 19th and Russell Force were hanging on and the superlative Petrol Company still clung grimly to the summit of Pink Hill, the last such bastion still in Allied hands. To the north the 18th paused and drew breath whilst Galatas itself pushed like an unsteady salient into no man's land at the apex.

Utz had decided to commit the 1st Battalion, which had escaped relatively unscathed, unlike his other units which had sustained heavy losses. Like the New Zealanders the Germans were feeling the pressure but these were, effectively, fresh troops and Utz was aware that if he offered the defenders a night's respite, much of his hard won gains might be jeopardised.

Schrank, in command of the 1st, was supported by a detachment of engineers from 3rd Battalion. He was to attack Russell Force from the south whilst the survivors of 2nd Battalion fought through the narrow streets of Galatas. Both sides had now committed all the forces available; it was do or die.

After a savage fight lasting half an hour the few remnants of the Petrol Company had been driven off Pink Hill, and Russell Force, as they fell back toward Galatas, discovered there were no friendly troops on their right, the whole position was imperilled. The line could no longer be held, the battle appeared lost but Inglis, after he'd heard of the fall of Wheat Hill, had cast around for any further reserves he could lay his hands on. What came to hand was Lieutenant Roy Farran and his two light tanks which had survived the earlier, unsuccessful attempt on Pirgos.

Farran, joined by a further two companies from the 23rd, was sent up to Kippenberger who flung the tanks into a charge through the streets of Galatas, now swarming with Germans. This bought some precious time; time to order the newcomers to prepare to retake the town at the point of the bayonet; no time for subtlety of approach, a straight dash forward behind the tanks and into the maelstrom.

When Roy Farran returned from his reconnaissance foray he reported the town 'stiff with Jerries'. In true Hussar spirit he was prepared to go again but already had two wounded crew. Volunteer replacements came forward from a group of detached engineers and these received a ten minutes crash course in armoured warfare before the attack was put in.

The orders for the assault were brutally concise:

D Company will be attacking on the left of the road, and we have two tanks in support but the whole show is stiff with Huns. It's going to be a bloody show but we've just got to succeed. Sandy, you will be on the right, Tex on the left. Now for Christ's sake get cracking.[46]

As the infantry slogged from the start line behind the rumble and clatter of the tanks, picking up momentum, now jogging towards the town, all sorts of individuals, carried away by the urgency of the moment fell in with them. Inevitably Captain Forrester of the Buffs, still without a helmet, was amongst these, at least one survivor from the valiant Petrol Company, happy for a chance to even the score, and a steady swell of men from the 18th and 20th, battered, bruised, bloodied but still undefeated.

Colonel Gray, still full of fight, had gathered a contingent from his battalion and would later recall how he would never forget the wild battle cries of the New Zealanders as they surged into the narrow streets, Farran's light tanks blazing away at their head.

In Galatas the Germans had thought that it was all over, at least for that night, and that they would continue to mop up in the morning. The storm that suddenly broke around them, terrifying in its suddenness, came as a very rude awakening. Before they realised it, the attackers were amongst them and a savage mêlée raged through streets and houses, the tanks hosing fire like demented metal monsters.

For a crazed interval, Roy Farran revved around the main square, firing at anything and everything whilst the infantry, winkling defenders from the warren of cellars, briefly lagged behind. The tank was drenched in fire, rounds smacking from the armour plating, tracer dancing like lethal fireflies. As the startled Axis called down a mortar strike, the percussion of bursting rounds swelled the concert of battle and a hit to the rear of the turret nearly propelled Farran clear of the vehicle.

The 23rd had been detailed to penetrate no further than the square but, finding themselves under fire, with the prospect of a clear cut victory and their blood white hot, they could not be held back. Again they charged:

> The consternation at the far side [of the square] was immediately apparent. Screams and shouts showed desperate panic in front of us and I suddenly knew that we had caught them ill prepared and in the act of forming up. Had our charge been delayed even minutes the position could easily have been reversed. By now we were stepping over groaning forms, and those which rose against us fell to our bayonets with their eighteen inches of steel entering throats and chests with the same hesitant ease as when we had used them on the straw packed dummies in Burnham. One of the boys behind me lurched heavily against me and fell at my feet, clutching his stomach. His restraint burbled in his throat for half a second as he fought against it, but stomach wounds are painful beyond human power of control and his screams soon rose above all the others. The Hun seemed in full flight. From doors, windows and roofs they swarmed wildly, falling over one another – there was little fire against us now.[47]

Throughout the battle for the streets of Galatas the fighting was savage, intense and, generally without quarter – Private D. Seaton took on a German machine gun, advancing steadily across open ground, using his Bren like a Tommy gun and firing from the hip. This cost him his life but comrades had worked around the gunner's flanks, secured by his covering fire and destroyed the nest with grenades.

Against seemingly insuperable odds the line had been restored. This epic charge and the wild, bloody mêlée in the lanes and dense packed houses, from the dark maze of ancient cellars to the fire swept killing ground of the square Galatas was, once again, in Allied hands. The fury of the counter stroke, tellingly supported by tanks, severely dented German morale. Utz's exhausted and demoralised troops were convinced this was not a purely local incident but the start of a general Allied offensive, the prospect of which had haunted them since the first cull of paratroops on 20 May.

>Our life style and instincts instruct us
>More cogently than any military precepts.
>Forward for New Zealand!

CHAPTER 8

One Large Stench

One final word about the reasons for the defeat. By common consent they were inferiority of land equipment and the enemy's practically undisputed mastery of the air. As for inferiority of land equipment, we might indeed have had more guns and more tanks in Crete if, like the enemy, we had been preparing for war for eight years; but only if we had had enough to be strong everywhere could we have been strong enough in Crete. But even if every aeroplane we had produced had been in the Middle East, we could not have got any greater fighter strength over the battlefield, and we could not have smashed the air-borne invasion in the air. Even with this colossal handicap, the issue of the battle hung in the balance for five or six days; and the course of the battle showed the enemy's best troops were no better than ours.[1]

Galatas had been a superbly fought action, driving the Germans from their gains in what appeared to be a moment of consolidation. Every Allied soldier who took part would know he had fought in an epic. It was, however, in strategic terms, no more than a delaying action. The place could not be held and Kippenberger recalled the bloodied and weary victors as, once again, a rain of German mortars began to pound the reeking streets and smashed houses.

Roy Farran was amongst the more seriously wounded who had to be left behind; for him the charge of the 3rd Hussars would end in captivity.[2] Despite the renewed ferocity of the bombardment the women and even children of the town sallied out into the shell-racked streets to aid the Allied casualties.

The survivors reeled back to take up positions along the Daratsos line, including the remnants of Russell Force and the gallant Petrol Company. Utterly exhausted, Kippenberger at last found his way to Inglis' makeshift HQ where his fellow officers were mostly gathered.

159

It was a sombre meeting, doubly galling in the aftermath of such
sustained heroism. Inglis mooted further counter-attacks; Dittmer,
arriving late, volunteered his superb Maoris but the die was already
cast.

Inglis knew that whilst an opportunity existed, created by the
reverse the enemy had just suffered, only an attack in force could
exert the necessary pressure to fracture his line. Puttick had not
attended the conference but sent Lieutenant Colonel Gentry who had
no fresh battalions to offer. He vetoed the Maoris going it alone.

Kippenberger described the course of this highly charged meeting:

> It was quite dark when we arrived at Brigade Headquarters and
> we stumbled around for some time among the trees. Inglis was in
> a tarpaulin covered hole in the ground, seated at a table with a
> very poor light. Burrows, Blackburn and Sanders were already
> there. Dittmer ... arrived a moment after me. It was clear to all
> of us that if this [counter-attack] was not feasible Crete was lost.
> It was a difficult operation, perhaps impossible: darkness, olive
> trees, vineyards, no good starline, only 400 men in the battalion.
> Dittmer said it was difficult; I said it could not be done and that
> it would need two fresh battalions. Inglis rightly pressed,
> remarking that we were done if it did not come off: 'Can you do
> it George?' Dittmer said 'I'll give it a go'. We sat silently looking
> at the map, and then Gentry lowered himself into the hole.
> Without hesitation he said 'No' – the Maoris were our last fresh
> battalion and if used now we would not be able to hold the line
> tomorrow. There was no further argument; it was quickly
> decided that Galatas must be abandoned, and everyone brought
> back to the Daratsos line before morning.[3]

The bitter truth was that only further withdrawal remained as an
option, swinging back to form a line with Vasey's Australians
presently sealing off the eastern flank of Prison Valley. In terms of
final defeat it was now a matter of when, rather than if.

Even now Wavell in Cairo was not fully aware that Freyberg had,
to all intents and purposes, decided to throw in the towel and concentrate on getting as many away as possible. In some measure this
was because the General himself had, until 26 May, continued to
paint a less depressing picture, suggesting that fighting for Maleme
was still raging and that the issue of possession remained in the
balance. Even as Farran's tanks roared into the streets of Galatas,

Freyberg had been cabling his superior to the effect that:

> Today has been one of great anxiety to me here. The enemy
> carried out one small attack last night and this afternoon he
> attacked with little success. This evening at 1700 hours bombers,
> dive bombers and ground strafers came over and bombed our
> forward troops and then his ground troops launched an attack.
> It is still in progress and I am awaiting news. If we can give him
> a really good knock it will have a very far reaching effect.[4]

However, before this could be sent an urgent dispatch arrived from
Puttick advising that the enemy had broken through at Galatas and
the line was no longer tenable. In consequence the General re-drafted
the final sentence of his own report to now read: 'I have now heard
from Puttick that the line has gone and that we are trying to stabilise.
I don't know if they will be able to, I am apprehensive. I will send
messages as I can later.' He then replied to Puttick advising that
whilst he understood how thinly spread the New Zealanders were
the new and shorter (Daratsos) line should be easier to hold and
must in fact be held.

On the 25th GHQ had scraped together enough Allied bombers to
mount a raid on Maleme. The sight of the RAF, commonly referred
to as 'Rare As Fairies' and even less complimentary translations,
gave a boost to sagging morale but, as the effort was so short lived,
the effect very quickly dissipated beneath the habitual weight of
relentless pressure from the omnipresent Stukas.

Even the might of the Luftwaffe proved fallible and the Allied
troops, hastily dug in to their new line, received a respite on 26 May
when a battalion from 85 Mountain Regiment, feeling their way
towards Perivolia, was repeatedly strafed by their own side. As a
result the Allied positions were not seriously troubled though
Freyberg was led to believe the 2 Greek Regiment, deployed around
Perivolia, was in difficulties. The Greeks stood south of 2/8th and
2/7th Australians whilst the shrunken New Zealand Brigade held the
line from Daratsos (Dittmer) to the coast.

As long as this present front could be maintained Souda Bay and
the hinterland of depots, stores, workshops, transport and logistics
could continue to function. This infrastructure required a near
division sized sprawl of non combat personnel who, although not
directly engaged, had been obliged to withstand the pernicious effect
of the endless air raids. No further ships were expected, the gauntlet

of the Kaso Straits was impassable, Cunningham would only permit limited nocturnal re-supply by fast destroyers which could complete a full turnaround under the shroud of darkness.

An attempt to land commandos at Paleokhora had been aborted but two companies were landed at Souda during the night of the 24th and the remainder of two weak battalions disembarked from destroyers *Hero, Nizam* and *Abdiel.* Layforce, as these 500 reinforcements came to be named, was under the command of Colonel Robert Laycock; 'A' Battalion was led by Lieutenant Colonel F. B. Colvin and 'D' Battalion commanded by Lieutenant Colonel Young. Laycock's intelligence officer was Evelyn Waugh, then a captain.[5]

It was not entirely accurate to say that there were no reinforcements available. 4 and 5 Brigades were indeed depleted but three crack battalions, 1st Welch, Northumberland Hussars and 1st Rangers, remained as Force Reserve on the Akrotiri Peninsula, largely uninvolved in the fighting since the first day. The difficulty was that these formations, whose presence in the battle for Galatas could have easily tipped the scales, came under Weston's rather than Puttick's command. When Inglis approached the General however, he got short shrift and Weston gave the impression that Force Reserve was to be kept under his own hand.

On the morning of the 26th Freyberg at last decided to divest Weston and place these men under Inglis to relieve the battered 5th. In the meantime Puttick had already reached the conclusion that the Daratsos line was untenable and a further withdrawal was necessary. When the divisional commander appeared at Creforce HQ to explain his reasoning he found the commander-in-chief had other ideas. Puttick was effectively superseded and all local forces in the Souda area would be placed under Weston.

Confusion ensued; when Weston failed to issue any direct orders for the night of 26/27 May Puttick approached him personally and found the General already abed and disinclined to cooperate. Puttick was told that as he appeared to have already decided on a further withdrawal there were no meaningful orders he, Weston, could issue. As a result a tragic development ensued – as the New Zealanders fell back, Force Reserve moved forward, confident in the belief it would have support on both flanks.

In consequence these three fine fighting units were left exposed to the full weight of the German advance, having driven what amounted to an untenable salient into what had, due to the New Zealanders' withdrawal, become enemy territory. This new defensive line lay along a sunken track that ran southwards from the western extremity of Souda Bay, and known as 42nd Street; so named because it had previously been home to 42nd Field Squadron Royal Engineers.

When Weston was finally made aware of the disaster about to enfold Force Reserve, he sent dispatch riders out to find 1st Welch and turn them around. It was too late. 'Custer's Last Stand', as a Welsh comedian had already dubbed the mission, was about to become something very similar.

The brigade passed through the ravaged and deserted streets of the capital encountering only a few exhausted stragglers. It was eerily quiet. They pushed on, confident they did have support somewhere on both flanks. By dawn it was obvious matters were deteriorating rapidly.

With patrols missing, the Welch and Rangers (the Hussars were forming the rearguard) were about to advance to contact with Ramcke's group moving eastwards along the coast and the 100 Mountain Regiment spilling down over the, now vacated, Daratsos Ridge. Worse, Heidrich's survivors, at last escaping from their own purgatory in Prison Valley, were about to cut off their retreat. Apart from numbers the Germans had an overwhelming superiority in terms of guns and mortars.

The result was never in doubt. The Welshmen fought hard but the odds were impossible. Force Reserve, by early afternoon, had ceased to exist as a fighting force. Of the 1,200 engaged, less than a quarter (mostly the Hussars) escaped. The final fighting formation available had now been used up, thrown away, in a dreadful muddle of command failure. The resulting, unseemly squabble amongst the senior officers involved, as to which of them was culpable, was nastily reminiscent of the wrangling at Balaclava in the Crimean War as to who had lost the Light Brigade. It was, of course, the waste, rather than the blame, which mattered.

Sir Lawrence Pumphrey, with the Noodles, was involved in the action which, for the Hussars who survived, heralded the commencement of yet another grim retreat. He was brought, by one of his troopers, a diary found on the body of one of the fallen *Fallschirmjäger*, as he was able to translate the German. He was touched by the dead man's account of his time in Athens, how closely this unknown enemy's excitement at seeing the Parthenon, matched his own.

Von der Heydte led his filthy and exhausted paratroops in their final advance toward Chania. The city fell without a shot, the scarecrow parachutists filing through the desecrated streets, fires smouldering in the charnel house air, the once elegant Venetian town reduced to a shattered ghost. Von der Heydte likened his scruffy troops to a band of medieval mercenaries. The Mayor, hastening to offer the city's surrender and spare his citizens further slaughter, at first refused to believe the tattered figure in front of him was indeed a senior German officer.

The pursuit of the Allies was left to Ringel's mountain battalions. The General assumed, not unnaturally, that Freyberg would continue to fall back eastwards and pick up the strong contingents still holding Rethymnon and Heraklion. As a result of this thinking, entirely logical, Ringel did not consider that the Allied survivors might rather seek a direct route to the south coast, there to await succour from the Navy. The option did seem unlikely, the road to the south was narrow, ill defined and vulnerable to attack from the air. The coast itself was known to be largely uninhabited with only a few and very small harbours.

Whilst the destruction of Force Reserve was in full play, Layforce was seeking orders. Both Laycock himself, with Waugh as IO, saw Colvin, who led the vanguard, then Freyburg and finally Weston, whose distraught appearance did not engender any degree of confidence. The commandos had arrived under a serious misapprehension, having been earlier led to believe the situation on the island was a good deal rosier than they now found. They had thought their mission, as befitted their swashbuckling image, would be to launch a series of spoiling raids against German lodgements in the west. Major F.C.C. Graham, writing only a few years after the end of the war, left a graphic record of their rude awakening on arrival:

No sooner had the ship anchored than boats from the shore began to come alongside and, just as the Brigade Commander, myself and other officers were bidding farewell to the captain of the minelayer, the door of the latter's cabin was flung open and a bedraggled and apparently slightly hysterical Naval officer burst in. In a voice trembling with emotion he said 'The Army's in full retreat. Everything is chaos. I've just had my best friend killed beside me. Crete is being evacuated'. Cheerful to say the least of it and something of a shock to the little party of commando officers armed to the teeth and loaded up like

Christmas trees, who stared open-mouthed at this bearer of bad news.

'But we are just going ashore,' I faltered.

'My God,' he cried, 'I didn't know that. Perhaps I shouldn't have said anything.'

'Too late now, old boy,' I said. 'You can at least tell us what the password is.' But he had forgotten it.[6]

While his commanding officer and IO were engaged on their largely fruitless tour of senior officers, Graham began to organise an initial deployment. The signs were not auspicious, the coast road was jammed with a chaotic muddle of formless troops falling back in disorder, 'dirty, weary, hungry – a rabble. One could call them nothing else.' The retreat was beginning to look like a rout.

Far to the west the defenders of Kastelli who had acquitted themselves so well on the first day had enjoyed a brief interlude of relative calm. Cut off from events to the east they could only await further developments. The fall of Maleme sealed their fate. By the 23rd patrols from 95th Engineers Battalion of the Mountain Division were probing their defences.

Coming across the bloated corpses of Muerbe's detachment left where they'd fallen in the earlier fight, the Alpine troops, under Major Schaette, concluded their fallen comrades had been mutilated after death by Cretan irregulars and this belief coloured their attitudes to those considered to be partisans. No quarter would be shown.

Next day the Luftwaffe paid the town a visit, sowing the seed of death and destruction through the streets. One bomb struck the gaol where Muerbe's survivors were still being held, which facilitated a mass breakout. The paratroopers took advantage of the confusion to raid Bedding's HQ and take him prisoner. This neat reversal prompted a swift attempt at rescue but this failed and cost Lieutenant Campbell who, with Lieutenant Yorke, had led the mission, his life.

As the bombers flew off the assault began. Artillery blasted the gallant defenders who, their ammunition exhausted, launched several spirited but doomed charges; some 200 fell. By the early afternoon much of Kastelli was in German hands. Bedding's 'B' battalion had dug in around the western flank of the harbour and held out grimly for a further two days of sporadic fighting, while the Axis guns systematically levelled the approaches. For a further four days, by which time the overall battle was irretrievably lost, the defenders clung on until the few survivors made good their escape into the sheltering hills.

Despite pleas from the captured Bedding, corroborated by Muerbe's survivors, that the civilian population had not indulged in any wholesale mutilation of German corpses, Schaette was determined that the Cretans should be made to understand the price of defying, and worse, humiliating the Luftwaffe. Over 200 male hostages were taken and summarily shot; thus did the people of Crete gain a first acquaintance with the brutal creed of Nazism. If the invaders thought that mere savagery and indiscriminate butchery could cow the islanders, then they were to be proved very much mistaken.

While the situation on Crete began, from 26 May, to unravel and Freyberg lost confidence in any prospect of hanging on, the demands from London for the island's continued reinforcement reached a new stridency. Churchill had, by now, become utterly fixated on a successful outcome and hurled fresh imperatives at Wavell, urgings that were not, in any way, rooted in the realities of a fast deteriorating situation.[7] On the morning of the 26th General Freyberg sat down to write the cable he must have been dreading:

> ... in my opinion the limit of endurance has been reached by the troops under my command here at Souda Bay. No matter what decision is taken by the Commanders in Chief from a military point of view our position here is hopeless. A small, ill equipped and immobile force such as ours cannot stand up against the concentrated bombing that we have been faced with during the last seven days. I feel that I should tell you that from an administrative point of view the difficulties of extricating this force are now insuperable. Provided a decision is taken at once a certain proportion of the force might be embarked. Once this sector has been reduced the reduction of Rethymno and Heraklion by the same methods will only be a matter of time.[8]

To Wavell and his colleagues at GHQ in Cairo it seemed impossible that the situation could have collapsed so comprehensively. In his reply the Commander-in-Chief urged Freyberg to consolidate, to hold on. Like General Ringel he assumed the Allies could fall back toward Rethymnon and Heraklion and bring up the reserves, largely unscathed, from both.

By now, however, Freyberg had already formulated and begun to implement his plan for a retreat southwards, over the mountains, to pray for evacuation from the tiny port of Sphakia. On the afternoon of the 27th he again cabled Wavell to admit the plan for a hoped for evacuation was already in hand.

The Commander-in-Chief now appreciated that the day was irrevocably lost and that all that could be done was to save as many as possible. He, in turn, cabled London to request confirmation that he should now proceed along this course. In the circumstances it was not possible to refuse. All thoughts turned to escape.

With a final bitter irony, the New Zealanders, on the morning of the 27th, afforded the Germans yet another reminder of their mettle. With Force Reserve outflanked and decimated, the 141 Mountain Regiment collided with the defenders of the 42nd Street Line.

The Maoris once again charged the advancing Germans who were taken completely by surprise:

> At first the enemy held, and could only be overcome by Tommy-gun, bayonet and rifle ... they continued to put up a fierce resistance until we had penetrated some 250 - 300 yards. They then commenced to panic and as the [Australian] troops from either side of us had now entered the fray it was not long before considerable numbers of the enemy were beating a hasty retreat.[9]

This spirited intervention left 121 of the Alpine troops dead on the field but nothing could now check the momentum of defeat which gathered like a leaden pall around the exhausted defenders. If gallantry had been enough then the outcome would have been a very different one:

> ... knew that I was taking part in a retreat; in fact I wondered if it should not be called more correctly a rout as, on all sides, men were hurrying along in disorder. Most of them had thrown away their rifles, and a number had even discarded their tunics, as it was a hot day ... Nearly every yard of the road and of the ditches on either side was strewn with abandoned arms and accoutrements, blankets, gasmasks, packs, kitbags, sun-helmets, cases and containers of all shapes and sizes, tinned provisions and boxes of cartridges and hand grenades.[10]

By the evening of 27 May Creforce HQ had evacuated Souda and was on its way to Sphakia. No possibility of stemming the tide of the

German advance now remained and Weston was left in local control. Laycock and Waugh were not alone in having formed a rather poor opinion of the general. Such was the magnitude of doubt that neither Hargest nor Vasey was prepared to await orders that might never come or would come too late.

Accordingly, with the fear of an outflanking move from the south, the surviving Australians and New Zealanders slipped away from the 42nd Street line during the evening, as their Commander-in-Chief and his HQ motored over the barren strip of the Askifou Plateau. The attack the Maoris had put in that afternoon had given the pursuers a sufficiently bloody nose and the withdrawal was not contested.

Another was to follow. The objective of the Allied withdrawal was the village of Stilos which lay on the northern approaches to the Askifou Plateau. As dawn broke on the 28th and the warm sun began its ascent toward the midday furnace, the Alpine troops renewed the chase.

Two companies of the 23rd, who were sleeping like the dead, had to be robustly roused by their officers when the German scouts were glimpsed. The New Zealanders crowded behind a dry stone wall, a perfect fire position and, as the Axis thundered on, began shooting at point-blank range. Many casualties were inflicted in the hot exchange and Sergeant Hulme earned his VC. Again the biters had been bitten.

While this action was being fought another engagement flared up further to the north, where the coast highway met the road heading at right angles southward toward Askifou, at Megala Khorion. Here the commandos, now dubbed Layforce, together with the Maoris, took on a mechanised column surging eastwards.[11] The 28th, as ever, gave a good account of themselves but the commandos, who comprised Republican remnants from Spain's civil war, serving with Young's 'D' Battalion, proved less enthusiastic.

Colvin's 'A' Battalion fared even worse. A number of the men were taken prisoner in the course of a couple of rather confused skir-mishes around Souda. Their commander's performance was so dismal Laycock sacked him and amalgamated the rump of the battalion with Young's.

As the 28th drew on into the afternoon, the long, normally languid hours where the great heat lay like a stifling blanket, the commandos who had now fallen back to a second line at Babali Hani came under renewed pressure. Layforce was stiffened by the 2/8th Australians and, later, by two Matildas from Heraklion. The late but timely arrival of the armour helped the composite force see off an attack put in by two of Ringel's alpine battalions.

Laycock, who'd narrowly avoided capture, managed to get through from Stilos to establish an HQ at Babali Hani. With due deference to their general's injunction that 'Sweat saves Blood' the mountain troops, once they'd probed a position, would then simply attempt to achieve fire superiority from the flanks, typically by seizing whatever high ground offered as a base for their machine guns and mortars. An eminently sensible tactical doctrine, and the mountainous terrain of the island provided ample scope.

With the Alpine troops edging around their flanks the Australians and commandos fell back, immobilising the two tanks which had run dry and a third which had appeared later. As light thickened in the thyme scented dusk, the sweet smells of the East Mediterranean spring tarnished by the tang of cordite, burning metal and rubber, the defenders slipped free to begin the long slog up toward the northern rim of the plateau. Ahead of them tramped the mass of the Allied survivors, ragged, weary, hungry and dispirited; an army that had coalesced into something very much resembling a leaderless mob, the stink of defeat hard upon them.

As a line of retreat, the road to Sphakia was far from being an ideal choice. From Vrysses the road climbed upward in a series of relentless hairpins, all signs of cultivation withered from the barren rock and scrub. There was no water, the stiffness of the climb a torment for a parched and exhausted man who stumbled along in a daze with his tattered footwear further disintegrating. The Stukas were frequent visitors, swarming like angry bees around the stricken carcase of the Allied Army:

> As we were retreating, the Stukas would come along and dive and strafe and we would disintegrate into the water table or into the bushes along the side of the road and I remember one morning this happened and after it was over we got out of the ditch very shaken and looking round to see if there were any killed or wounded, and two hard-boiled soldiers Bill and Les. And Les said to Bill: 'My God, I prayed for you when the Jerries were strafing us and you were in that ditch,' and Bill said: 'Sorry Les, too late, I prayed this morning.' And all the others looked around and thought that if old Les prayed there must be room for everybody else to pray.[12]

Lawrence Pumphrey, struggling up the hard, steep ascent, was finding the going murderously tough. His own boots had been with the local cobbler on 20 May and he'd been marching in a pair of desert boots he'd found chucked in a ditch. He soon realised why the previous owner had cast these aside so disdainfully. By now his feet were badly blistered and each step was agony. He made it as far as Sphakia only because his brother, John, managed to liberate a motorcycle with sufficient fuel still in the tank.

Bad as the roaring Me109s were, there were actually less of them; the relentless timetable for Barbarossa had denuded the available aircraft – Russia's impending misery at least partially spared the survivors on Crete as they wound their grim and weary passage up the mountain.

Although the rearguard and the fighting troops were able to maintain their morale, the less cohesive units, British and Middle Eastern support personnel, were disintegrating, plundering such dwindling stockpiles of foodstuffs as could be found. The Australians and commandos, being the last, thus fared badly. Freddie Graham, the brigade major, performed miracles of logistics but even so, hunger began to bite.

Kippenberger saw the padre, heavy laden with water bottles which he carried to sustain his flock, his own lips dry and cracked. Other survivors remember the endless, cruel succession of hairpin bends, each one promising the summit, only for the corner to reveal yet another, uphill and ahead. Vehicles which ran out of fuel or broke down were ruthlessly trundled over the edge to crash, tumbling and spinning, down the steep, rock strewn slopes. Graham recorded his own view of the march:

> The road was jammed with troops in no formed bodies shambling along in desperate haste. Dirty, weary and hungry, they were a conglomeration of Australians, a few New Zealanders and British, and Greek refugees. They had only one thing in common and that was a desire to get as far as possible from Canea – a rabble one could call them, nothing else ... desperately we pushed our way on to the road and tried to push past the motley throng which straggled all over it. All day the sky was thick with enemy aircraft in many cases flying at only a few hundred feet and every now and again coming down to bomb and machine gun the troops trudging along the road. All day the stream of the retreat flowed steadily but wearily on. When enemy

aircraft approached the bulk of the men tried to scatter off the road or hid in the ditches; some impervious to threats such as: 'Lie down you bastard or I'll fucking shoot you,' bore steadily on.[13]

Desperate for food the men resorted to foraging, some living like outlaws in caves and abandoned farmsteads; the Spanish Republicans, survivors of three brutal and hungry years of bitter civil strife, proved particularly adept, even inviting Laycock and Waugh to an impromptu feast. When men did lie down to snatch a few hours or even minutes of rest, they sank into such a stupor of exhaustion they could not, even by the most robust methods, be roused.

While the bulk of the Allied forces were beginning the gruelling ascent toward the plateau and, from there, down the steep crevice of the Imbros Gorge to Sphakia, the garrisons at Rethymnon and Heraklion remained largely unaware of the unfolding disaster in the west. They, for their part, had acquitted themselves triumphantly: the lodgement at Rethymnon had steadily been eroded and the German survivors further east around Heraklion were depleted, demoralised and completely isolated without the consolation of having attained any of their objectives.

It came as a singularly rude awakening then when Brigadier Chappell summoned his battalion commanders to a conference at his HQ on the morning of the 28th and advised them that they and their men were to be taken off by ship from the port that evening. Total secrecy was to be maintained and the Greek survivors who had fought so valiantly alongside the Allies were to be left in ignorance – there were simply not enough boats available.

This was galling for the British and Australian officers, to abandon these brave Allies to the summary justice of the *Fallschirmjäger*. One of the most popular officers in the Black Watch, whose companies had fought Brauer to a standstill, Major Hamilton, who had vowed that 'the Black Watch leaves Crete when the snow leaves Mount Ida' ,had become a casualty that very morning.

As Chappell and his HQ staff set about the dispiriting business of burning and destroying, one of Pendlebury's partisan contacts appeared and made a passionate appeal that the Allies hand over any equipment they could; there was no rebuke for the abandonment. The brigadier was deeply moved by the man's courage and bearing, and promptly complied. Everything that could be used was given to the Cretans.[14]

That evening the troops, astonished that they were now to flee from a field they believed won, went about the grim chores of disabling guns and vehicles, burying ammunition stocks, destroying and contaminating anything that might succour the invaders. In the event the evacuation proceeded in text book fashion. Silently the battalions filed down toward the port, leaving behind all of the familiar landmarks they had thought safe. The city was in little better state than Chania, the bombing had devastated large areas, with the stink of raw sewage and the pervading reek of decaying corpses heavy in the night air.

Patrick Leigh Fermor[15] noticed what appeared to be a nervous, rather boyish soldier who turned out to be a Cretan sweetheart. He simply decided to look the other way. This young lady was not unique for a number of others and a few civilians managed to get aboard the rescue vessels. These were under the command of Rear Admiral Rawlings with the cruisers *Orion* and *Dido*[16], escorted by half a dozen destroyers. In all the ships took off 3.486 troops; the entire garrison.[17]

Inevitably, small groups from outlying posts and the wounded crowded into the hospital at Knossos were left behind, only hearing the news of their abandonment from locals. Some simply waited for captivity, others prepared to cross the mountains to reach the south coast and some took to the hills with Cretan partisans.

It was now the early hours of 29 May after a flawless evacuation but the destroyer *Imperial* had suffered undetected damage during the earlier bomb attacks and her steering gear jammed completely. *Hotspur* was instructed to take off the complement of crew and soldiers. This too, was carried out with great efficiency but a small cadre of Australians who had earlier sought solace in liquor were by now utterly insensible. When the stricken ship was sunk by torpedos, they went to the bottom with her.

As the heavily laden *Hotspur* struggled to catch up, the remainder of the squadron was obliged to cut their speed to no more than 15 knots. In a grim replay of the previous action at sea the first bright rays of daylight found the ships north of the Kaso Strait. Almost immediately the dreaded siren wail of the Stukas pierced the clear morning air; an overture for six hours' relentless hammering from above.

Orion was hit twice and bracketed by half a dozen near misses.

Bursting through the armour plate, the bombs tore into the bowels of the vessel and exploded amongst the densely packed evacuees, nearly 300 were killed outright in the inferno and as many more injured. Ironically it was those who'd volunteered to chance the deck and man Bren guns that escaped.

> At first light the Stukas started their dive bombing. With hundreds of others I was packed like a sardine down in a mess deck. The enemy pressed home their attack and *Orion* was hit twice. The second bomb came down between decks and there was indescribable horror among the hundreds of men there. I blacked out from cordite fumes in my lungs and must have been unconscious for some hours. I owe my life to my pal Frank Humphrey who hauled me to safety after the stair had been blown away and just in time before the watertight doors were closed, sealing that part of the ship.[18]

Dido was also struck with similar results as a bomb penetrated the crowded canteen, leaving another 100 men dead. Some wounded survivors were drowned as the area was flooded to contain the flames. *Hereward*, too had become a casualty. It was evening by the time the damaged flotilla steamed, battered and gaping, into Alexandria; almost 20 per cent of the garrison, taken off without loss, had been lost in the sea chase.

The situation which unfolded at Rethymnon was somewhat different. Campbell had succeeded in eliminating or severely shaking the two German lodgements to the east and west of his sector; that aside, matters had been reasonably stable. Like Chappell he had no immediate appreciation of just how bad things were generally and the Naval officer, commanding a supply drop from Souda on the 28th, carried no fresh orders. Lieutenant Haig could, however, intimate that he was ordered to make for Sphakia and that some kind of a general retreat southwards appeared to be in the offing.

In fact no orders reached Campbell. He had not completely destroyed the German positions to the west and ruled out a concerted attack to re-open the coast road; both of his tanks had been written off in the attempts. His orders were to hold the airstrip; that he had done, most admirably, and would continue to do. By now, however, Wittman's Advanced Guard was steaming eastwards, the 85 and 141 Alpine Regiments were following on and, at last, Ringel could call on full armoured support – tanks from 31

Armoured Regiment had been successfully disembarked at Kastelli.

This powerful battle group, far stronger than anything the Allies could now scrape together, was to relieve firstly Rethymnon and then Heraklion. As previously noted it was this entirely logical appreciation that had allowed the tattered tail of Freyberg's army to turn southward, largely unmolested, but it spelt the end for the garrison at Rethymnon. It was only around dusk on the 29th that Campbell heard the shocking news that Heraklion was abandoned. Worse still, German patrols were already probing the town of Rethymnon and ranging shells from their artillery were beginning their drumming dirge. That night the Australians held sad vigil by the beach, flashing torches into the darkness in the hope that the Navy had vessels lying offshore but none came. Next morning the pressure was swiftly increased. Campbell was clinging to the aerodrome with the 2/1st – Sandover commanded the 2/11th. Evidence of the German strength was all too obvious; the outposts were already being driven in.

Campbell came to the conclusion that there was little point in going on and that there was no option but the bitter pill of capitulation. Sandover disagreed. He believed that anyone prepared to take the chance should be given leave to take to the hills. In the event he, with a gallant band of thirteen officers and thirty-nine other ranks, did just that and, after many adventures, they were rescued from the south coast by submarine.

At 8.30 a.m. on 29 May Campbell and the rest surrendered, plodding through the ravaged landscape and marching into captivity with the added humiliation of being made to pull the trailers containing enemy equipment:

> The fields on each side were sprinkled with dead and no-mans-land near the town itself was even more thickly spread with corpses ... We saw a paratrooper, still attached to his parachute hanging from the telephone wires. Half his head had been blown off.[19]

Having mopped up resistance at Rethymnon, Wittman forged westwards to Heraklion where Brauer's survivors had emerged to take control of the ruins. The charge was spent, he sent a fast flying column of motorcyclists toward Aghios Nikolaos. Their mission was to link up with the Italians. Determined to play his part in the tragedy of Crete, Il Duce had sent amphibious forces to occupy the eastern extremity of the island around Sitia on 28 May. These fresh invaders, landing from the Dodecanese, were not engaged after the battle was safely won. Italian gunboats had launched some half hearted spoiling attacks on Royal Navy vessels and the port of Ierapetra had been bombed on the 25th.

Lawrence Pumphrey, having managed to reach the beach, was amongst those taken prisoner. His brother John, who'd acted as a movement officer in the final stages of the evacuation, found himself a passenger in a Sunderland flying boat, but his brother officers and troopers were left behind.

With his lacerated feet too painful even to hobble, Lawrence was driven into captivity on the back of a German truck. Like so many Allied prisoners he found he had little personal animosity toward his captors. They, for their part, behaved well. Food was scarce but, after a few days on the north side of the island, he was flown back to Athens, a very different journey from the first. He was amongst a group of British prisoners, temporarily held in an old Greek barrack pile at Salonika. The buildings were so infested it was deemed wisest to sleep out on the parade ground. Some weeks later the prisoners were herded into cattle trucks for the long journey north. At Belgrade they received welcome succour from the Red Cross. At Lübeck they were put into a newly constructed camp. Lawrence was suffering from jaundice and shivering in his battered tropical kit. Later that year he was moved to a vast compound at Warburg, situated on the cold, bare expanse of the Westphalian Plain. He remained a year in Warburg where, with others, he planned to escape by means of a tunnel. By this time officers and men had been segregated, though he still had the company of a handful of fellow Noodles who'd all been captured on Crete. In June 1943, he was part of a mass escape from the camp at Eistedt in Bavaria. Sixty-five prisoners tunnelled clear of the camp but were subsequently all re-captured. Lawrence Pumphrey finished the war in Colditz.

Despite these continuing reverses and the collapse of the Allied position, the Cretans attached no blame to the individual soldiers whom they continued to help, despite the inevitability of murderous reprisals. As Sandover and his party were about to take their leave by submarine, one of the local volunteers said this to him: 'Major, my greatest wish is that you will take a glass of wine in my house the day we are free. That is all I wish to live for.'[20]

No more eloquent testimony could be desired.

CHAPTER 9

The Navy Must Carry On— Evacuation 28/31 May

Admiral Cunningham belonged firmly to the Sea Dog tradition of Hawkins and Drake but he also cared passionately for both the men and ships under his command. The devastation which the air-sea encounters had wrought upon his tired fleet affected him deeply:

> I shall never forget the sight of those ships coming up harbour, the guns of their fore-turrets awry ... and the marks of their ordeal only too plainly visible, [*Orion* was] a terrible sight and the messdeck a ghastly shambles.[1]

The choice was a stark one; to send these battered ships out again to attempt to rescue the bulk of the Crete garrison from Sphakia, was fraught with great risk. The fleet simply could not withstand another round with the Luftwaffe. The relatively minor evacuation from Heraklion had already shown, in all too graphic terms, what aerial bombardment could do to crowded vessels; what horrors might then accrue on the larger scale.

For the thousands who would be waiting on the beaches, this could only mean surrender – the loss of the entire garrison. It was a prospect too dreadful to contemplate. From Corunna to Dunkirk the Navy had always achieved the miraculous. It had been done again in Greece and must now be attempted from Crete. It might, as the Admiral observed, take years to construct a battle fleet but 300 years of tradition, once lost, could never be recovered. The operation would proceed.

South from the Askifou Plateau the road now swings around a dizzying round of hairpin bends as it drops nearly 2,500 feet to the narrow coastal plain, the great bulk of the mountains rearing up behind, crowding the ribbon of shore. In May 1941 this road did not exist, it had not been finished, coming to an abrupt halt at the mouth of the Imbros Gorge where it spilled from the rim of the plateau.

The gorge is truly daunting, a narrow, precipitous canyon of broken rock and scrub that opens like an abyss, as dry as sandpaper and hot as a flaming cauldron. It was this descent which now confronted the unbelievably weary survivors, nor was there time to draw halt or seek a respite; the dogs of war were hot on their trail.

During the hours of darkness on the night of 30 May the Germans began to move across the plateau; by dawn they were at the mouth of the gorge. The 'Saucer', as Askifou was dubbed, was defended by the two Australian battalions, supported by the final trio of light tanks, all that remained of the injured Roy Farran's original detachment. The pass itself was held by the 23rd.

Colonel Utz, with the final, clear-cut victory in sight, was not inclined to take undue risks, 'sweat saves blood' as, for the Germans, did the Luftwaffe. The demands of Barbarossa had, as noted, denuded the available squadrons so pressure from the air was less overwhelming than before. The Axis troops had pilfered large amounts of discarded Allied tropical kit and had even plundered civilian garments, including ladies' underwear, to form makeshift headscarves.

True to Ringel's doctrine Utz opted for outflanking moves, seeking to find a way around the heights at the head of the pass and thus cut off the defenders holding the mouth. One prong of this attack was thwarted by the adverse topography but the second element, guided by a collaborator, succeeded in almost cutting the pass.

Brigadier Inglis immediately replied by sending in Lieutenant Upham's company from the 20th. A confused fight developed among the wilderness of rock and wild rhododendron, the Germans were checked and driven back with loss – one Kiwi was reputedly dangled by the calves to fire his Bren around a difficult corner.

If the fighting troops were still full of fire, the scene at Sphakia was considerably less edifying. Thousands of exhausted and ragged men, all discipline and cohesion gone, crowded the narrow ribbon of coast, living like vagabonds or wreckers in the many limestone caves that riddled the hillsides. They were without arms or equipment, leadership or control, a rabble of bacaudae or bouche inutiles.[2]

During the night of 28/29 May the Navy made its first run to the south coast. A trio of destroyers sailed at full throttle to take off 744 men, including many of those wounded. This operation was accomplished without loss. The RAF had pledged a maximum effort to provide adequate cover and, though the friendly planes missed the flotilla and despite the attentions of prowling Ju88s, no casualties were sustained.

Next evening a far larger formation, Force 'D', commanded by Rear Admiral King, and comprising cruisers *Phoebe* and *Perth*, two anti-aircraft cruisers, *Coventry* and *Calcutta*, together with a quartet of destroyers and the landing vessel *Glengyle*, set sail for Sphakia.

As he was steaming north, the tattered survivors from Heraklion limped into port and Cunningham, having conferred with Wavell, who in turn spoke to Tedder and Blamey to glean the airmen's view, signalled Whitehall to advise of the continuing losses. He warned of still greater loss should more vessels, particularly the vulnerable *Glengyle*, be lost.

Should further efforts be limited to destroyers only? In spite of the prevailing climate of despondency Cunningham was prepared to try the last and risk all. The Admiralty attempted a compromise suggesting *Glengyle* be recalled. By then (the signal was not received till nearly 8.30 p.m.), it was too late, Force D was committed in its entirety. To assist, Cunningham dispatched a further three destroyers.

That night some 6,000 men were taken off, an amazing feat given the host of difficulties – the boats from *Glengyle* proved their worth as ferries. The Luftwaffe did put in an appearance, Perth suffered a hit to her boiler room but, at least in part, due to the RAF shield, the air attacks proved neither as sustained nor as damaging as before. More and more German planes were being withdrawn from the sector to meet the timetable for Barbarossa and this relentless siphoning of aircraft undoubtedly helped to relieve the pressure.

On the 30th another flotilla, this time made up entirely of destroyers *Napier*, *Nizam*, *Kelvin* and *Kandahar* steamed out of Alexandria. These ships, under Captain Arliss were less fortunate. *Kandahar* developed engine trouble and *Kelvin* was damaged by bombs; both were obliged to return. Nonetheless Arliss proceeded with the two remaining vessels and, using the boats left by *Glengyle*, lifted a further 1,400 from the beaches; no mean achievement.

Exhausted, filthy and ragged survivors marvelled at the calm, clean efficiency of the Navy who provided food, hot coffee and medical facilities. They had been rescued from a squalid world of defeat and confusion, indiscipline and disorder into what seemed an unreal haven. As Dr Stephanides observed wryly: 'even the officers' uniforms were neat.'³

The ships slipped away in the pre dawn but, despite an RAF escort, were pounced on by dive bombers, screaming down out of the morning sun. *Napier* was hit and for an anxious while lay dead in the water. Kippenberger, who was on board, suddenly felt the thought of safe deliverance might have been premature:

> [he] felt a stunning concussion ... everything loose in the cabin crashed all ways, and I found myself sitting on the floor in darkness. My first thought was that the cable announcing my safe arrival would not now be sent.⁴

Despite the fury of the attack and amid the rattle of the Oerlikons, spattering rifles and Brens, the crew managed to get the stricken vessel under way and she staggered safely into port. Kippenberger's men, spruced up and shaven, disembarked in an orderly manner. Their country had every right to be immensely proud of them.

Another who departed, this time by flying boat that same night, was General Freyberg who left Weston in command of the beaches. This was a very difficult moment for a commander who cared deeply for his men, racked by guilt that elements of the New Zealanders must inevitably be left behind. The fact that he and his staff had knowledge of ULTRA intercepts meant their feelings in the matter were of no account, their capture must be avoided.

Hargest led the remains of the 5 Brigade down the narrow defile from Askifou on the morning of the 30th – the contrast between the fighting troops and the rabble they encountered was startling. All pretence of military bearing and formation had now gone, to the extent the Kiwis had to set up a steel tipped line of pickets to deter the mob from rushing the ships.

Priority was given to the formed infantry though not all would be accommodated – HQ units and officers, the necessary core of any unit, were taken off first then as many NCOs and other ranks as could be accommodated. Such was the press of *bouches inutiles* that the final rearguard from the 28th stood with their arms, including sub-machine guns, at the ready.

Despite the damage sustained by Arliss' flotilla, Cunningham once again sent Rear Admiral King with cruisers *Phoebe*, *Abdiel* and a brace of escorting destroyers on the morning of the 31st. This last effort netted another 4,000 safely taken off that night. *Calcutta*, which had been subsequently deployed to provide anti-aircraft cover did not return.

Subsequently, there was some bitterness that all of the senior officers appear to have been got safely off: 'One of the worst episodes in that affair was the notion that superior officers were specially valuable, that there was an obligation on them to save themselves... .'[5]

A case in point is the controversy which has arisen over Colonel Laycock. The commandos, who had done an excellent job as part of the rearguard, were the last to descend to the beachhead and Freyberg, prior to his departure, issued orders that they would be the last fighting troops to embark. Although this was confirmed by Weston on the 31st, Laycock then advised these orders had been amended as he still had two full battalions in Egypt, so that he and his HQ, including Waugh, got off.

In a personal memoir Waugh appears to suggest that Laycock had arranged the matter after a private talk with Weston – the General decided that for the commandos to be captured would be a greater loss than the unfortunate Colvin who was elected to stay in his place.

Graham, Layforce's Brigade Major, was dispatched by his colonel to attend Weston and Colvin in what remained of Creforce HQ. There Graham was ordered to write down the orders for capitulation which detailed the luckless Colvin to proceed at first light and surrender to the first German detachment. Weston, leaving the two junior officers with a spare bottle of gin and a reserve fund of 1,000.000 drachmae, then left his HQ to be taken off by flying boat.

If Graham thought this spelt the end for him he was mistaken, for Laycock and Waugh appeared very soon afterward and gathered all available Layforce personnel for evacuation. Clearly Laycock had no intention of being left behind. Despite the rather dubious provenance of his orders, the prevailing climate of *sauve qui peut* dictated that all who could escape, did.

A degree of confusion arose over the situation of Young's detachment strung out in the hills covering the approaches, along with the marines and the 2/7th Australians. In the event, although advised of the possibility of evacuation that night by Waugh's batman Private Ralph 'Lofty' Tanner, Young did not make the beachhead. As Colvin appears to have slipped away with Laycock, Waugh, Graham and the rump of Layforce, the unenviable task of negotiating the final capitulation fell to him.

Though the commandos appeared to have jumped the queue, the 2/7th Australians filed down to the beachhead, a difficult and frustrating journey as they had a long march through the crowded night. Their disappointment can only be imagined when they found themselves stranded and that the last of the ships had departed. They had fought hard and well throughout and, until this bitter moment, had no thoughts of throwing in the towel; indeed they were inclined to seek orders to fire on those who were already laying out white flags.

As he set off on his search for a German to surrender to, leaving his adjutant, Michael Borwick, to break the news to the men[6], Young encountered Colonel Walker, commanding the 2/7th and, as he was senior, passed the poisoned chalice to him. Walker walked alone toward the village of Komithades where he offered his surrender to an officer from the 100 Mountain Regiment.[7] The Battle for Crete was at last over.

In fact it was not quite over. As a group of Allied soldiers left behind on the beach, believing hostilities now at an end, tried to get a cooking fire going they attracted the attention of an Me109 which, as the Luftwaffe presumably had not yet been informed of the capitulation, opened fire. The strafing killed one man and filled a wounded sergeant with another dozen bullets. German soldiers who tried to attract the plane's attention were also shot up.

Though most seemed to accept the reality of surrender with resignation, others were traumatised:

The realisation was stupefying, dumbfounding. In all my previous existence and I had then had nearly 35 years of it [never] had I received news that knocked me all of a heap as this had.[8]

I have never felt so terribly as I did at that moment. In fact, I don't think that I had ever really felt ill at all till then. Any troubles I had in the past were mere ripples compared with this tidal wave; I was deeply disappointed; I felt frustrated and shamed – above all ashamed.[9]

Amongst those waiting for captivity two groups in particular had reason to be especially apprehensive – these were the commandos generally and particularly the Spanish Republican element. Hitler tended to favour shooting special forces and so everyone prudently divested themselves of their knuckle duster daggers, 'fannies' as they were called. The Spaniards' medical officer, Captain Cochrane, came up with the inspired notion that the men should pose as volunteers from Gibraltar.

Not all were willing to give up and stories abound of desperate and heroic attempts to find small boats and chance the hazard of the 200 mile journey to North Africa. Such a group discovered one of *Glengyle*'s boats abandoned near the port, which they pulled under cover.

On the night of 1 June they began their attempt. Hit by German fire and later damaged when it ran ashore, the flimsy craft continued to float, though ten volunteers had to be left behind to improve buoyancy. After the fuel was exhausted they rigged a sail which was soon flapping listlessly in dead calm. Food and water dwindled, but on 8 June they sighted land and they eventually drifted to landfall near Sidi Barrani.

Despite the success of Admiral King's final effort, Cunningham had decided that this was all that could be done. London continued to press for one last effort but the Admiral was unshakeable:

> I was forced to reply that Major General Weston had returned with the report that 5,000 troops remaining in Crete were incapable of further resistance because of strain and lack of food. They had, therefore been instructed to capitulate, and in the circumstances no further ships would be sent.[10]

This was clearly the correct decision; the Navy had kept its promise.

CHAPTER 10

Glad to have seen the day— Occupation and Liberation 1941–1945

An island with as long a history of occupation and revolt as Crete was bound to have developed an instinctive belief in merciless treatment for traitors. Collaborators knew they could expect no mercy if caught. One German agent captured by andartes begged to be allowed to commit suicide. They broke his legs with heavy stones some way from the edge of a cliff so he had to crawl the rest of the way to push himself over.[1]

The Germans anticipated that they could cow the population simply by a ruthless application of brute force and largely indiscriminate violence. They did not begin to understand the character of the people whose island they were presently occupying. The Cretans knew all about occupiers; theirs was a long history of occupation and resistance to the occupier. Every man they shot sparked another vendetta.

Crete was a pyrrhic victory for the Axis, the battle cost them a total of 6,580 dead, missing and wounded; of these 3,352 were dead and a large proportion of the loss was borne by the *Fallschirmjäger* – as Student later admitted: 'For me ... the Battle of Crete ... carries bitter memories. I miscalculated when I suggested this attack, which resulted in the loss of so many valuable parachutists that it meant the end of the German airborne landing forces which I had created.'[2]

In this he was correct. Hitler was appalled at the scale of loss and concluded that the day of the paratrooper was over. Student's dream of vertical envelopment was a further casualty of the fight. The dust had barely settled when all eyes were turned eastwards and to the invasion of Russia. With Barbarossa under way, the eastern Mediterranean and Middle East slipped down the strategic agenda.

For the rest of the war the island was a backwater – as a crowning irony it was perhaps beneficial to the Allied cause in that the garrison which, at its height, comprised some 75,000 men, tied down a large number of Axis troops.

No grand strategic design followed from the capture; the idea of moving on to attack Cyprus disappeared in the wake of Barbarossa. The function of the battle in Hitler's eyes had been, in part, to add to the deception surrounding the overall build-up for the attack on Russia and to provide a secure back door in the Mediterranean. Beyond this the Führer had no interest in further operations.

The idea of deception was played out by Goebbels and even Goering, who informed a conference of his own Luftwaffe commanders in Paris that Merkur had been a full dress rehearsal for a reinvigorated invasion of England. ULTRA, of course, informed the British to the contrary and Bletchley was aware of the reasons for the build up along the Soviet frontier. The Russians, however, proved disinclined to accept the validity of the subsequent warnings.

None of this was of particular interest to the Cretans who were solely interested in driving out the invader. In the course of the struggle some 3,474 islanders were killed, many of these in the course of the final German entrenchment around Chania when hostages were murdered almost as a matter of course.

Once the main Axis forces had retreated from mainland Greece in 1944, the island's shrunken garrison was effectively abandoned and withdrew to a fortified enclave around the capital. Here they survived in a state of siege until the last gasp of the Reich when Admiral Dönitz, as Hitler's successor, formally ordered their capitulation in May 1945.

To accept the surrender of the German governor, the RAF flew him to Heraklion – the aircraft landed initially at Maleme; a highly symbolic reversal. David Hunt, who had been with Creforce HQ four years earlier, recorded the moment with some understandable satisfaction:

> It was an agreeable example of the wheel turning full circle. At that time the Germans were all concentrated at the west end of the island, and our main concern was to keep the Cretans from falling on them. The solution was to move them all into the Akrotiri Peninsula, where I had watched the gliders landing four years earlier, and put a cordon of British troops across the neck. From there they were taken away by ships and sent to Souda Bay.

I was glad to have seen the day.[3]

General Kurt Student was one of those German officers charged with war crimes. He was probably never an ardent Nazi but clearly empathised with the National Socialist dream of a strong Greater Germany and was, at best, prepared to play along to achieve his career goals. Many others, of course, fitted in with this acquiescence.

Student's trial took place on Lüneburg Heath in May 1947 and, whilst he was acquitted on the majority of counts, was convicted on others. There has to be a question of the soundness of the verdict for there was no evidence of him having specifically ordered acts which would constitute war crimes. Still plagued by the effects of his old head injury, the architect of Germany's airborne elite was subsequently freed on medical grounds and sank quietly from view.

It was upon the necks of his former subordinates that the axe fell most heavily; three of the island's military governors, Andrae, Muller and Brauer, were put on trial for their lives in Athens. The first escaped with life imprisonment when the sentences were handed down in December 1946 but the other two were both hanged on 20 May 1947 – symbolically on the anniversary of the battle. This was felt, certainly in Brauer's case, to be both distasteful and unfair, smacking more of retribution than due process.

The Battle for Crete effectively ended Wavell's active career as it had done Student's. His position, already weakened by reverses in the Western Desert, became untenable in the wake of the failure at Crete. Churchill was only seeking an excuse to rid himself of Wavell whom he clearly disliked, and the victor of Beda Fomm found himself shunted sideways into obscurity.

Public opinion in the USA seemed largely unaffected by this further Allied reverse and Vichy continued to glower in hostile neutrality. There was an inevitable backlash from the Antipodes. In Australia the disaster hastened the government's fall and, from New Zealand, Prime Minister Fraser journeyed to London, demanding a fuller explanation as to why the Dominion troops had been left with what was now perceived as an impossible situation. Before doing so, on 7 June, he penned a highly critical cable:

Operations in Crete seem to have been largely the result of chance. The driving from the Greek mainland of various forces (including New Zealanders) with different degrees of equipment but on the whole ill supplied and to some extent disorganised, with an embarrassing number of refuges, seems to have found them on the island, which it was then decided to hold. As you know we had no previous knowledge that it was intended to defend it, and it seems clear to me now that the island was, in fact, indefensible with the means at Freyberg's disposal against the scale of attack which eventually developed. It seems to me also that it should have been as clear before the decision to defend Crete, as it is now that troops without adequate air protection (which it was known could not be provided) would be in a hopeless position, though it is obvious that the scale of the German air attack was larger and more intense than was foreseen.[4]

Freyberg became the main target for his political master's wrath. This was blatantly unfair, for the General had been placed in an impossible position, his loyalty split between the normal chain of command and the demands of the home government. For a professional soldier such as he, there could be no question of refusing Wavell's orders to accept command of Crete; that would have been unthinkable and rightly so. The full position should have been disclosed to the Dominion government by the War Cabinet whose stance was, at best, misleading.

The meeting became very heated with Fraser exhibiting the full wisdom of hindsight:

No matter who your Commander-in-chief or what his rank may be, [the Prime Minister thundered] it is your duty to keep us in touch with the situation ... when you are ordered to take part in operations you will personally find out whether there is air cover for operations anticipated and you will communicate with us and tell us you are satisfied; and secondly your troops will not be exposed without tank support to hostile tank attack.[5]

As a sop the Dominions were granted an additional, if largely token, presence in the councils of the War Cabinet but two precepts were identified and agreed:

(1) No further operations of a similar nature were to be mounted unless the Dominion troops taking part could be guaranteed adequate air cover and (2) they would not be expected to take on mechanized opponents with nothing more than 'their rifles and their courage'.[6]

The defeat had actually cost the Allies some 1,750 soldiers killed, as many wounded, and over 12,000 marched into captivity. The Royal Navy lost something in the order of a further 2,000 men during the course of the various battles at sea. For that vast, motley horde left on the shore at Sphakia, capitulation meant a further, grinding march back up the steep, broken defile of the Imbros Gorge, over the furnace bowl of the Askifou, the long descent to Vrysses and, eventually incarceration in a makeshift POW pen on the site of the No. 7 General Hospital. And here they languished. Conditions were primitive, supplies irregular and sanitation uncertain. The stench of decaying flesh clung, cloying and sickly, to the heavy, heat-laden air. From the surrounding districts sporadic bursts of fire echoed as the Axis 'dealt' with suspected partisans. Other troops, deployed as the Sonderkommando von Kuhnsberg, tracked Allied survivors and escapees hiding in the hills.

Despite the savagery of reprisal, many locals risked the German bullets to bring food to the prisoners, a number of whom, inevitably, succumbed to dysentery. In time most were transported, firstly to the mainland and then to camps far away to the east. In defiance of convention some 800 were forced to work on the repair of Maleme airstrip.

The story of the resistance on Crete is a heroic one. A series of rather swashbuckling characters such as Paddy Leigh Fermor, W. Stanley Moss, Jack Smith-Hughes, Xan Fielding, Tom Dunabin and Ralph Stockbridge worked with local guerrilla bands in conditions of great hardship and danger. Coming in by caique and submarine, the raiders teamed with partisans for selective sabotage and intelligence gathering. Offensive operations were fraught with risk, not just for the initiators but also the populace who could expect no mercy from the murderous ruthlessness of the occupiers.

Pendlebury had died during the fight for Crete; others followed. Mike Cumberlege was captured in 1942 off the mainland coast and, after three years in the dreadful confines of Flossenburg concentration camp, was executed with three others only four days ahead of the liberation.

Gunner D.C. Perkins, whom the Cretans christened Captain Vassilios, escaped from Galatas prison camp and, with a fellow NCO, Tom Moir, crossed to the south coast to look for a boat. The pair were assisted and sheltered by locals. Moir succeeded in getting away in a stolen boat; Perkins had to wait to be taken off by submarine. Both returned, Moir to look for other escapees – after a number of exploits he was finally recaptured – Perkins, now trained in sabotage, attached to Xan Fielding's cadre. From July 1942 the pair were firmly established in the White Mountains, working with local guerrilla bands. An attempt at a concerted rising, orchestrated by Mandli Bandervas, proved abortive and brought down a hurricane of reprisals. Perkins reformed the survivors into a company-sized unit which operated from the village of Koustoyerako.

Perkins, Captain Vassilios, became the stuff of local legend, leading his andartes in a dazzling series of raids and ambushes. On one occasion the partisans surrounded a German patrol of twenty men who barricaded themselves in a stone hut. Perkins went forward and winkled them out with Mills grenades. Half the Germans were killed in the attack, the rest were simply shot out of hand. For all the romantic dash this was a bitter war of close quarters and no mercy. Perkins himself was wounded in the spine by a German bullet – the local butcher acted as surgeon, without the benefit of anaesthetic.

Operations on Crete were supported by a host of buccaneering small boats and by air drops of arms and ammunition. The Allies and andartes lived hard in the arid mountains in cold winters and baking summers. They went in fear of betrayal by German sympathisers, of whom, even on Crete, there were many, and in the knowledge their actions could bring down fearful wrath on innocent civilians. One of the effects of this close cooperation was to avoid the pernicious polarisation between partisans from the left and right which scarred the resistance effort on the mainland, and lit the fuse for the bitter civil strife that followed the German withdrawal.

The purpose of the resistance was primarily intelligence gathering, sabotage and low intensity operations. There was not the intention to arm the populace for a general rising. This was a prudent and highly effective policy for the andartes were able to tie down a very large number of Germans and their despised Italian allies. Although the Cretans were intensely anti-royalist they were equally opposed to Communism and the left was never able to gain an effective toehold. As a result the island was spared the two years of murder and strife that wasted the nation as a whole.

Perhaps the most celebrated guerrilla action on Crete, and that which guaranteed lasting fame for both W. Stanley Moss and Patrick Leigh Fermor, was the spectacular abduction of the island's governor, Major General Heinrich Kreipe. A veteran of the Russian front, the General must have viewed his posting to the island as something of a 'cushy number'. The preferred, initial target had been the detested General Muller who had a particularly savage reputation.

On 26 April 1944, as the General's gleaming staff car swept from the impressive gates of his official residence at the Villa Ariadne, Moss and Leigh Fermor, both dressed as Wehrmacht, were waiting to flag down the vehicle. Both Kreipe and his driver were abducted; the latter being surplus was, subsequently, quietly done to death. With one of the kidnappers at the wheel the British drove confidently through the streets of Heraklion, the governor's official pennants deterring any check.

The car was abandoned on the coast road toward Rethymnon with a scattering of documents in English and a note to the Germans advising the abduction was solely the work of British agents and officers from the Greek Government in exile. This may not have entirely eliminated the threat of reprisals as a number of villages were torched around this time, possibly as a consequence of separate actions. On 14 May at 11.00 p.m. General Kreipe was taken off by boat to begin his captivity.

When, in February 1944, Perkins himself fell in an ambush, his sorrowing followers buried him. His grave, so far from home, became something of a shrine and a photograph, from 1951, carried the following note:

> Grave of the most fearless of fighters ever to leave New Zealand, known to all Cretans as the famous Kapitan Vassilios. Killed over 100 Germans single handed during the occupation. Led a guerrilla band, and fell from machine gun fire in February 1944, near Lakkoi – the last gallant Kiwi killed in Crete. This man is honoured by all Cretans.[7]

A man could wish for no finer epitaph – and indeed this could be applied to all those men who had come so far to fight and die for freedom. Although the battle might be lost, such sacrifice was never in vain.

CHAPTER 11

Remembrance

The New Zealanders and other British, Imperial and Greek troops who fought in confused, disheartening and vain struggle for Crete may feel that they played a definite part in an event which brought us far-reaching relief at a hingeing moment.

Winston Churchill.

On 29 September 1945, 100 officers and men of the New Zealand Division, including their commander General Freyberg, attended a memorial service on Crete. For the three days the party remained on the island they were lavishly feted by locals. As early as June that year the site for the Commonwealth War Graves Commission (CWGC) cemetery, to be located in Souda Bay, had been decided.

The final resting place for those who died in the fight for Crete was designed by Louis de Soissons, the architect of, *inter alia*, Welwyn Garden City and who was also to be responsible for the design of all CWGC cemeteries in Greece, Italy and Austria. One can traverse the globe and all of the many hundreds of CWGC graveyards are instantly recognisable; the white Portland stone markers, bearing name, unit, number, rank and date of death beneath a regimental crest of those, most poignantly, 'known unto God'.

All British and Commonwealth cemeteries exude the calm of an English country garden, the perfect rows of headstones aligned as though on final parade, verdant with flowers and meticulously tended grass. That at Souda, located three miles east of Chania in the cusp of the Akrotiri Peninsula, contains 1,509 burials dating from the fighting, nineteen from the Great War and thirty-seven others moved from the former consular plot in 1963.

The cemetery overlooks the placid waters of the bay, still used by Greek warships. The design is symmetrical with a total of sixteen plots, the memorial standing centrally. The forecourt is attractively paved in limestone and marble, patterned with smooth, rounded pebbles. Many of those who lie there are marked as unknown – during the occupation the Germans moved the impromptu battlefield burials to four large grave pits at Chania, Galatos, Rethymnon and Heraklion. Those who have no known grave are commemorated on the Allied War Memorial at Phaleron Cemetery, Athens.

In the flush of their pyrrhic victory, the Germans constructed a memorial, latterly surrounded by modern tourist development, on a mound west of Chania, showing a diving eagle grasping a swastika, the badge of the *Fallschirmjäger*. Below is a stone plinth commemorating those who died in the struggle. It speaks well for the tolerance of the Cretans that this odious memento was left standing in 1945. It has, apparently, since fallen into disrepair.

An official German war cemetery was constructed on the slopes of Hill 107 and opened on 6 October 1974. There is a marked difference in the appearance of German as opposed to Allied sites, most noticeable, not here, but on the Western Front. There is none of the calm melancholy of the country garden; more Wagner than Mozart, dark flat plinths and an absence of planting, a brooding, Teutonic sadness.

On Hill 107 there is a total of 4,465 graves divided into four plots, each commemorating one of the four main areas of conflict. Olive groves struggle down the western flank to the dry banks of the Tavronitis and, from the summit, one can clearly see the airstrip at Maleme below, the strength of the position is immediately clear.

Beneath the low walls of each plot, each stone tablet records the names of two who fell, the 300 whose remains could not be found are remembered on the memorial. At the foot of the hill the entrance portico houses a small exhibition. It is a fitting place and one of the early superintendents was none other than George Psychoundakis who, at the request of the Association of German Airborne Troops, brought Brauer's remains from their war cemetery in Athens.

The inscription on the memorial plaque reads:

In this graveyard rest 4465 German dead from the war years 1941 - 1945. 3352 of them died during the battle of Crete between 20 May and 1 June 1941 ... They gave their lives for their Fatherland. Their deaths should always make it our duty to preserve peace among nations.

There are other remembrances – an RAF memorial at Maleme, one

for the Royal Artillery on the Akrotiri Peninsula, the Stavremenos
monument to the Cretan resistance and a plaque at Prevelli
Monastery which records:

> This region after the battle of Crete became the rallying point for
> hundreds of British, Australian and New Zealand Soldiers, in
> defiance of ferocious German reprisals suffered by the monks
> and native population. They fed, protected and helped these
> soldiers to avoid capture and guided them to the beachhead
> where they escaped to the free world by British submarines.

In the Naval Museum in Chania which overlooks the lovely Venetian
harbour, there is an extensive exhibition charting the history of the
battle through models, excellent and numerous photographs, maps,
displays of uniform, small arms and equipment. Further information
is housed in the archives here, those in the Chania History Museum
and at the Historical Museum in Heraklion.

For those travelling to Sphakia the War Museum, run by M.
Hatzidakis and his son, at Askifou is not to be missed – a wonder-
fully eclectic collection of battlefield relics, expended munitions,
machine guns, sub-machine guns, rifles and pistols, an MP40 and a
Thompson next to an ancient percussion survivor pressed into
service by the resistance.

For the battlefield tour it is probably best to begin at Maleme with
the German cemetery, the airfield is still very much in use by the
Greek military and perceived trespassers are not likely to be made
welcome. The airstrip is, in any event, best glimpsed from the
summit of Hill 107. The iron bridge over the Tavronitis still stands
adjacent to its modern replacement and is accessible on foot.

There is a war memorial in the square of Galatos by the church,
and a further monument to the Greek troops in Prison Valley below.
The whitewashed walls of the gaol still stand, as uninviting but
prominent as they were in 1941, the corner by the Greek memorial
affords a good view looking north of, on the left of the road, Pink
Hill and, on the right, Cemetery Hill.

Moving eastwards along the coast to Rethymnon, where the proud
citadel of the Venetian Fort still sits proudly, it is possible, driving
along the old coast road toward Stavromenos, to ascertain the
location of the air strip and to look inland towards the rise, on the
left, of Hill A with Hill D to your right.

The journey uphill from Vrysses to Askifou and then down the precipitous road to Sphakia is spectacular but not for the faint-hearted. The experience is a 'must' for those who really seek to understand the closing stages of the battle. The difficulties imposed by the topography are immense and it leaves one with a sense of awe that the Navy, in such conditions as obtained, succeeded in taking off so many of the garrison.

Crete has, of course, changed very considerably since 1941. The development is most apparent along the north coast, particularly west of Chania where the old settlements have been linked by a continuous, unrelenting strand of modern tourist sprawl; an endless procession of trendy condominiums, hotels, restaurants, supermarkets and shops. Those who fought there in the course of the battle would struggle to recognise the townships they garrisoned.

Nonetheless, there is much that has not changed. The Venetian harbours at Chania and Rethymnon appear timeless, the jumble of narrow alleys behind the port, now populated with boutiques and bistros, resonates with the cultural legacy of conflicting occupations.

The history which is marketed to the booming tourist industry focuses on the splendours of the Minoan Age and the palace culture, the events of 1941 are a largely forgotten footnote, no coach borne hordes descend on the Allied or German cemeteries, the survivors a steadily dwindling band. Crete is perhaps too remote to participate in the vogue for battlefield tours that has engulfed the Western Front and Normandy.

None of these means the Battle for Crete was or is irrelevant. For the Germans their pyrrhic victory tasted little better than defeat. The British, Australian and New Zealand troops took on the best of Nazi Germany and, man for man, fought them to a standstill.

Perhaps the last word should be left to Kurt Student, the architect of vertical envelopment who, in 1944, was left to witness from the ground as the vast aerial armada spearheading Operation Market Garden flew overhead. He is said to have remarked how magnificent it would have been to have, at least once, possessed such power.

The wheel had come full circle.

Notes

Chapter 1

1. Garret D., The Campaign in Greece and Crete, HMSO, 1942, p5.
2. Quoted in Simpson A., Operation Mercury – The Battle for Crete London, 1981, pp268-69.
3. ibid., p139.
4. Henry James Colossus of Maroussi, 1941.
5. Das Inselmeer der Griechen, 1962.
6. Rethymnon is a Venetian survivor, the great fortress still commands the headland and the teeming, crowded lanes of the port cluster below; the tall Italian tenements studded with distinctive Ottoman shuttered balconies, the overlapping layers of occupation. A single minaret still gives a hint of vanished Islam.
7. Clark A., The Fall of Crete, London, 1962, p1.
8. ibid., p2.
9. ibid., pp2-3.
10. Quoted in Simpson, op. cit., p22.
11. ibid., p24.
12. For an assessment of the Intelligence War refer to Appendix 3. The Germans, sure of the infallibility of Enigma, chose to account for their intelligence failures as having emanated from their Italian allies due to the latter's (a) general incompetence or (b) deliberate treachery.
13. Simpson, op. cit., p28.
14. ibid., p29.
15. Clark, op. cit., p6.

Chapter 2

1. Garnett, op. cit., pp5-6.
2. Major General R.F.K. Belchem quoted in Simpson, op. cit., p49.
3. OKH = Oberkommando des Heeres – Army High Command.
4. OKW = Oberkommando der Wehrmacht – overall command of the German armed forces.
5. Quoted in MacDonald C., The Lost Battle: Crete 1941, London, 1993, p51.

6. Quoted in Simpson, op. cit., p50.

7. Quoted in Clark, op. cit., p9.

8. The campaign in the Western Desert opened with a British offensive in December 1940. Wavell had correctly divined that the battle would best be waged by armoured forces spearheading a lightning attack. His plans were entirely successful and Graziani's more numerous forces were at first defeated then pushed into rout, tens of thousands of Italian prisoners were taken.

9. OKW had estimated that Barbarossa would take some ten weeks to achieve a successful outcome for Germany.

10. The MNBDO was intended, as its name implied, to act as a mobile, self contained force, able to fully provide AA cover, coastal and ground security for any bases occupied on a short term basis by the fleet. It also boasted a landing and base maintenance detachment. Wavell had decided to send MNBDO to Souda on 2 April when the anchorage was to be upgraded to full base capacity in support of the Greek intervention.

11. Quoted in Simpson, op. cit., p52.

12. Colonel de Guingand, quoted in Clark, op. cit., p16.

13. Quoted in Simpson, op. cit., p63.

14. ibid., p75.

15. Quoted in MacDonald, op. cit., p57.

16. ibid., p57.

17. George Orwell, writing in *Partisan Review* July 1941 Collected Essays, Journalism and Letters of George Orwell Vol. 2 My Country Right or Left? London, 1970, p148.

18. Digger's doggerel quoted in MacDonald, op. cit., p148.

19. Second Lieutenant Upham quoted in Simpson, op. cit., p86.

20. Quoted Simpson, op. cit., p91.

21. There was a whiff of treachery in the air as it was suggested certain elements within the higher echelons of the Greek forces had already been in tentative contact with the Germans to discuss terms for surrender.

22. V. Ball quoted in Simpson, op. cit., p99.

23. ibid., p110.

24. Garnett, op. cit., pp41-2.

25. Clark, op. cit., p20.

Chapter 3

1. Paratroops' marching song.

2. The Ju52/3 mg7e transport was the workhorse of the German

forces, manufactured near Dessau, the distinctive three engined
B.M/W. 132T radials, 830 hp each monoplane performed a similar
role to the DC-3 in the Allied armies. It could carry eighteen troops
at a maximum speed of 189 mph with a range of 930 miles.

3. Quoted in MacDonald, op. cit., p5 – a similar anecdote attaches
to Student's future opponent Major General Freyberg who, when
his command post was strafed, stood unconcerned in the road
whilst his less blasé staff dived for cover! 'Interesting isn't it?' he
was said to have remarked to his driver, who was unsure if the
general was offering an observation on the amount of bullets
kicking up around him or the reaction of his officers!

4. The Deutsches Forschunginstitut fur Segelfug (DFS) 230A glider
could carry eight paratroops or 2,720 lbs. of equipment, its
wingspan was a huge 72 feet with a length of 37 feet, normal
towing speed was 131 mph. It must have been an eerie sight indeed
to see the great broad winged aircraft swooping silently down.
They were, of course, of lightweight construction and horribly vulnerable to ground fire.

5. The Germans had developed an airborne version of the 28 mm
PzB 41, which featured a tapering barrel (28 mm - 20 mm) this
gave the tungsten steel shot a muzzle velocity of 4,600 ft. per
second. The German troops were also equipped with a 75 mm
RCL gun – a Recoilless Rifle. This had a standard 75 mm round
but the shell case had a frangible plastic base which held for long
enough to allow pressure to build up, start the projectile moving,
then blow out through an aperture in the breech block to create the
balancing pressure of gas, necessary to create a 'recoilless' effect.

6. Lieutenant D. Davin quoted in Simpson, op. cit., p24.

7. ibid., p123.

8. In the course of his post war captivity and interrogation Student
complained that his head wound was causing him severe physical
and emotional trauma; particularly, depression, speech impediment
and a desire to shun all human contact. This could be dismissed as
an attempt to avoid the allegations of war crimes relating to the
treatment of British POWs on Crete which were being levelled
against himand in respect of which he was later acquitted.
Nonetheless it would seem the General did suffer lasting impair-
ment as a result of his injuries. This did not, however, shorten his
life as he survived into old age and died at eighty-eight!

9. British Paratroop marching song, quoted in Simpson, op. cit.,
p123.

10. Friedrich August von der Heydte came from a family of
Bavarian gentry; a career soldier who served with both infantry and
cavalry before pursuing an academic career as a lawyer at
Innsbruck University. A devout Catholic he rejoined the ranks on
the outbreak of war, transferring to the Paratroops in August 1940.
For his service in Crete he was awarded the Knight's Cross. He had
an active war and was finally captured in the winter of 1944
during the Ardennes Offensive. He was a member of the anti-Nazi
clique within the German armed forces and was fortunate to escape
the consequences of a possible involvement in the July Plot.
Another officer with a very similar name was targeted by the
Gestapo in error! He later returned to the quieter reaches of university life.
11. Julius 'Papa' Ringel was a career soldier in the Austrian Army,
an experienced officer of the old school whose oft quoted maxim
was 'sweat saves blood'. A dedicated Nazi, he commanded the 5th
Division from its inception in October 1940. Ringel did not see
eye-to-eye with Student, whom he regarded as a dreamer. Student,
for his part, saw his colleague as a plodder. It was his leadership of
the German campaign in Crete that determined the latter stages,
something Student would bitterly resent.
12. Wolfram von Richthofen, a relative of the famous 'Red Baron',
commanded Fliegerkorps VIII which conferred equal rank with
Student who, therefore, had to request rather than order, as
doubtless he would have preferred. Von Richthofen was an
acknowledged master of the aerial element of blitzkrieg. He had
also flown fighters with distinction in the Great War and had the
reputation of being something of a prima donna.
13. Bernard Ramcke was tasked with the difficult job of preparing
the mountain troops for their unexpected, airborne role. This will
not have been a popular appointment. He was very much the
fighting soldier, who'd fought through the trenches in the previous
war and with the uncouth Freikorps; thereafter, he took over
Meindl's command after the General had succumbed to wounds.
Despite being in his early fifties his vigorous actions during the
campaign brought him a Knights Cross to which was later added
the Oakleaves, Swords and Diamonds, (only twenty-seven of this
much coveted decoration were awarded in the course of the war).
14. The Italians had been responding to pressure to mount
offensive naval operations from their German allies – in part this
had contributed to the disaster off Cape Matapan on 28 March,

1941. In relation to Crete the Italians were to claim the Luftwaffe discouraged any Italian naval presence in the vicinity, as their pilots 'had never flown missions at sea before and were unable to distinguish between friendly and enemy ships,' (quoted in MacDonald, op. cit., p73). They concluded that the Germans sought to prove that naval operations could be successfully discouraged by air power alone.

15. Prison Valley (actually the Ayia valley) was so named due to the presence of a large whitewashed gaol at the eastern end of the shallow depression. This structure, which still stands, was overlooked by the hills south of Galatas, see Beevor A., *Crete; The Battle and the Resistance*, London, 1991, p112.

16. Quoted in MacDonald, op. cit., p83.

17. Winston Churchill quoted in Simpson, op. cit., p119.

18. Clark, op. cit., p22.

19. MacDonald, op. cit., p115.

20. ibid., p115.

21. ibid., p122.

22. ibid., p126.

23. ibid., p127.

24. ibid., p134.

25. ibid., p134.

26. For information on ULTRA please refer to Appendix 3.

27. Quoted in MacDonald, op. cit., p135.

28. Freyberg outwardly remained resigned about the defeat on Crete and though he was never again to hold an independent command, he continued in senior combat roles during the war. He never wrote his memoirs and remained silent until his death so it is difficult to review events from his personal perspective.

29. Quoted in Clark, op. cit., p26.

30. ibid., p26.

31. ibid., p28. Both Australian and New Zealand formations were comprised entirely of volunteers, unit designation are prefixed with '2' – this denotes that this was the second Australian volunteer force, the first having been that which served with such distinction in the previous war. The same applied to the New Zealand contingent, again this was the 2nd Expeditionary Force.

32. The tanks sent to Crete comprised the Matilda Infantry Tank and Mk. VIB Vickers Light Tanks. The former was powered by twin diesel motors and mounted with a 2-pounder gun (or 3 in. howitzer) and a Besa 7.92 mm machine gun. The Matilda, which

weighed in at 25 tons, was heavily armoured and had achieved some successes, particularly in the early battles of the war, around Arras, in 1940. It was, however, both very slow and under gunned; the design of the turret ring inhibited the fitting of a larger calibre weapon. Its earlier title of 'Queen of the Battlefield' was already faded when these tanks were deployed on Crete. The Vickers was essentially a tracked reconnaissance vehicle mounting both light and heavy machine guns, a small boxy hull riding on a Horstman Suspension system (also employed in the design of the Bren Carrier). The Vickers was obsolete by 1941, its armour too thin, its firepower inadequate.
33. Quoted in Clark, op. cit., p29.
34. Extract from an unpublished memoir of the Northumberland Hussars.
35. Simpson, op. cit., p139.

Chapter 4

1. 'E.F.U.' in *New Zealand Expeditionary Force (NZEF) Times* 18 January 1943, p5.
2. Major H.G. Dyer quoted in Simpson, op. cit., p152.
3. Simpson op. cit., pp152-3.
4. ibid., p153.
5. ibid., p153 – the first airborne wave comprised 280 bombers, 150 dive bombers, 180 fighters, 500 Ju52 transports and 70 to 80 gliders, towed by Ju52s.
6. ibid., p155.
7. ibid., pp154-5.
8. ibid., pp158-9.
9. ibid., p160.
10. ibid., p156.
11. ibid., p155.
12. ibid., p157.
13. ibid., p169.
14. Many reports speak of the descending parachutists being riddled with bullets in the air. In fact a man descending by parachute is an extremely difficult target and most fatalities probably occurred as the troops hit the ground; here, they were at their most vulnerable. The Germans were also severely hampered by the design of their parachutes which relied on the static line. Once released and in the air the individual soldier could not, unlike their Allied counterparts, control the direction of their descent.

15. ibid., p162.
16. ibid., p156.
17. ibid., p170.
18. ibid., pp154-5.
19. ibid., pp162-3.
20. Quoted in Clark, op. cit., p62
21. MacDonald, op. cit., pp179-80.
22. ibid., p180.
23. ibid., p180.
24. Quoted in Clark, op. cit., p64.
25. Simpson op. cit., p165.
26. ibid., p166.
27. MacDonald, op. cit., p180.
28. Simpson, op. cit., p167
29. Clark, op. cit., p64.
30. ibid., p64.
31. There were suspicions attaching to the loyalties of the Greek governor of Ayia Gaol and Heydte recounts this official presenting himself and offering his services in any capacity.
32. Simpson, op. cit., p164.
33. ibid., pp164-5.
34. Garnett, op. cit., p32.
35. Quoted in Clark, op. cit., p91.
36. Simpson, op. cit., p178.
37. ibid., p178.
38. ibid., p177.

Chapter 5

1. Quoted in Clark, op. cit., p93.
2. Quoted in Simpson, op. cit., p178.
3. Count von Blucher came from a long and distinguished line of soldiers, his most famous ancestor being Wellington's great ally at Waterloo. Two of his brothers also fell in the fight for Crete, (Simpson, op. cit., p215).
4. Bruno Brauer had served with distinction in the trenches winning the Iron Cross, 1st and 2nd class. He was appointed to the command of the 1 Parachute Regiment in 1938 and added the Knights Cross to his laurels during the battles in the Low Countries. He became overall governor of Crete in 1943 before being sent back to the Russian Front in the last, hopeless days of the war. His execution for war crimes took place on the sixth

anniversary of the battle, though many considered the conviction to be politically motivated and unsound.

5. Simpson, op. cit., p176.
6. ibid., p176.
7. Garnet, op. cit., p14.
8. Major General Freyberg, quoted in Simpson, op. cit., p179.
9. Dr. H. Neumann quoted in MacDonald, op. cit., p202.
10. OL 2170.
11. OL 2/302.
12. Beevor, op. cit., p91.
13. ibid., p91.
14. *Gefechtbericht XI FL Korps – Einsatz Kreta.*
15. Beevor, op. cit., p91.
16. Keegan J., Intelligence in War, London, 2003, p196.
17. Simpson op. cit., p174.
18. Simpson, op. cit., p174.
19. Lieutenant Colonel Andrew had won his VC for outstanding gallantry during the First World War. A career soldier, he had risen from the ranks; like Hargest he was an officer of proven courage. It was like his superiors, his inability to grasp the concept of vertical envelopment that affected his decision making and, in the crucial fight for Hill 107, he was badly served by the chain of command above.
20. James Hargest was a native New Zealander. He served with distinction during the Great War and, between the wars, farmed and became involved in politics. Physically fearless, he looked very much the ruddy faced, solid farmer that he was. In his early fifties in 1941 it is not unlikely he was suffering from acute exhaustion at the time of the battle. He was killed on active service after D-Day in 1944.
21. Simpson, op. cit., p185.
22. ibid., p190.
23. ibid., p190.
24. Clark, op. cit., p72.
25. ibid., p74.
26. Major General K.L. Stewart CB DSO (at the time Brigadier CGS to NZ Division), quoted ibid., p78.

Chapter 6

1. Report of the US Naval Observers and journalists to US Military Attache in Cairo quoted in MacDonald, op. cit., p245.

2. Gerd Stamp a JU 88 pilot quoted ibid., p249.
3. Cunningham's own nephew was amongst the dead.
4. Oswald Janke quoted in MacDonald, op. cit., p242.
5. Janke again quoted, ibid., p239. The British initially claimed to have inflicted 4,000 casualties but in fact many who were forced to abandon ship were later rescued and, due to the *Lupo*'s heroic stand, only the leading elements of the flotilla were destroyed, the rest managed to disperse. The actual loss appears to be just over 300, though all of the heavy equipment went to the bottom and no succour reached the Germans on Crete by sea.
6. Ios, one of the most magical of the Greek Islands and said to be the spot where Homer died.
7. Cunningham records a final conversation with *Gloucester*'s captain, H.A. Rowley, as she was preparing to leave Alexandria. He was very concerned about the condition of his crew, ground down by incessant action at sea. Cunningham never saw Rowley again, his body was washed ashore west of Mersa Matruh some four weeks after the battle, recognisable only by the uniform and documents in his pockets. 'It was a long way round to come home.' See Clark op. cit., p117.
8. MacDonald op. cit., pp146-7.
9. Clark op. cit., p119.
10. ibid., p120.

Chapter 7

1. General Freyberg quoted in MacDonald, op. cit., p203.
2. Quoted in Simpson, op. cit., pp202-3.
3. Geoffrey Cox was not properly an IO – his main job, which he'd continued in spite of the invasion, was to produce the *Crete News* – he came upon the German order more or less by accident when he called in at Creforce HQ having previously assisted in the clashes on the Akrotiri Peninsula. He discovered the order lying in a pile awaiting dispatch to Cairo and began to translate; this he achieved with the aid of German pocket dictionary!
4. General Student quoted in Simpson, op. cit., p203.
5. Simpson, op. cit., p204.
6. ibid., p204.
7. Clark, op. cit., p107.
8. ibid., p107.
9. Simpson, op. cit., p205.
10. ibid., p205.

11. Clark op. cit., p123.

12. ibid., p123.

13. ibid., p122.

14. Simpson, op. cit., p214; Lieutenant Colonel Walker, before he left the meeting, briefed his second in command, Major Marshall, that the men would need to be ready to move by dusk (around 8 p.m.). Consequently he felt all the transport should be assembled at least three hours beforehand. This was no mean task; the constant attention of the Luftwaffe made movement of vehicles in daylight a highly fraught business. Many drivers had been so shaken they abandoned their trucks to take shelter at the first intimation. Nonetheless Marshall proved equal to the considerable challenge and succeeded in rounding up a good number of vehicles.

15. ibid., pp216-17.

16. ibid., p217.

17. ibid., pp217-18.

18. As recalled by Colonel Dittmer quoted in Clark, op. cit., p131.

19. ibid., p133 – the platoon officer in question was, in fact, Lieutenant Upham.

20. ibid., pp219-20 – W.O.L. Young.

21. ibid., pp219-20.

22. ibid., p220.

23. ibid., p221.

24. ibid., p222.

25. ibid., pp223-24 – R. Wenning.

26. ibid., p224 – 'Lofty' Fellows.

27. ibid., p240.

28. ibid., p241.

29. ibid., p241.

30. ibid., p242.

31. ibid., p242.

32. ibid., p242.

33. ibid., p243.

34. Clark op. cit., p138.

35. Heidrich went on to command the re-named 1st Parachute Division in the Italian campaign, where he remained until captured by a patrol from the 3rd Battalion Grenadier Guards. A story is told that whilst under interrogation by the battalion IO, Nigel Nicolson, the pair began a discussion as to the merits of German and Allied small arms. Heidrich, ostensibly to make a point, asked the sentry to hand over his Thompson sub-machine gun. The

soldier was about to oblige before Nicolson stopped him. Heidrich 'merely smiled'.

36. Clark op. cit., p141.
37. Simpson op. cit., pp249-50 – Lieutenant N. Hart.
38. Clark, op. cit., p142.
39. ibid., p149.
40. ibid., p151.
41. Simpson, op. cit., p251.
42. ibid., p252.
43. ibid., p252.
44. ibid., p232 – Corporal T. McGee .
45. Clark, op. cit., p252
46. ibid., p145 (from the New Zealand Official History) .
47. ibid., pp156-7 – Lieutenant Thomas.

Chapter 8

1. Garnett op. cit., pp62-3.
2. As the wounded Farran, crippled by a gangrenous knee injury, was awaiting transfer to Athens he was taunted and spat upon by a group of Italians, erstwhile prisoners of the Greeks. A German sentry, clearly outraged by this abuse of a helpless, stretcher bound casualty, set about the chief tormentor, giving the Italian 'one of the most severe beatings I have ever seen administered in my life.' See MacDonald, op. cit., pp295-6.
3. ibid., p161.
4. Quoted in Clark, op. cit., p159.
5. ibid., p159.
6. So far Layforce had no reason to celebrate its stay in the Middle East theatre; a bewildering catalogue of operations had been proposed and then aborted. The commandos had earned the unhelpful nickname of 'Belayforce' on the troopship *Glengyle* some wit had scrawled 'never in the history of human endeavour have so few been buggered about by so many'. See Beevor A., op. cit. p195.
7. ibid., p195-6.
8. An attempt had been made to send the 2nd Battalion the Queen's Regiment, embarked on *Glenroy*, but aborted due to severe air attacks. Ten Hurricanes were dispatched to Heraklion, inevitably, too few and too late – the first half dozen were destroyed by friendly ground fire, two turned back and disappeared, the remaining pair did land but were shot up on the ground! It was also proposed to send a squadron of Bristol

Beaufighters but the logistical difficulties and the realisation that this limited intervention could not hope to influence the outcome killed the notion. See Clark, op. cit., p164.

9. ibid., p164.
10. Captain Baker, quoted in Simpson, op. cit., p263.
11. This force was Advanced Guard Wittman and comprised: 95th Motorcycle Battalion; 95th Reconnaissance Unit; detachment from 95th Anti-Tank Battalion; elements of Motorised Artillery and Engineers.
12. Keith Elliot quoted in Simpson, op. cit., pp267-8.
13. ibid., pp268-9.
14. Antonis Grigorakis Satanas of Kroussonas, one of Pendlebury's resistance captains, a redoubtable fighter later evacuated by caique to Alexandria – he had developed symptoms of the cancer that would kill him.
15. Patrick Leigh Fermor, then serving as Chappel's IO, a noted Grecophile who had tramped overland across Europe before the war, was to prove a distinguished member of the cadre of British officers who assisted with the resistance and, with W. Stanley Moss, participated in the capture of General Kreipe (see Chapter 20). Since the end of the war he has become a distinguished writer and traveller; a swashbuckler in the Pendlebury mode.
16. *Ajax,* damaged in the course of an earlier air attack, had limped back to Alexandria.
17. Beevor, op. cit., p209.
18. Private J. Renwick quoted in Simpson, op. cit., p260.
19. L. Lind quoted in MacDonald, op. cit., p280.
20. ibid., p283.

Chapter 9

1. Quoted in MacDonald, op. cit., p285.
2. *Bouches Inutiles* – literally 'useless mouths' – the expression seems to date from medieval siege warfare when civilians, unable to contribute to the defence, were expelled from walled towns; in the case of the Hundred Years War the besiegers would often forbid the refugees to pass through the lines so they were left in the hostile limbo of no-man's-land.
3. Quoted in Simpson, op. cit., p289.
4. ibid., p290. Anthony Beevor records that on the first night of the evacuation *Napier* took off 36 officers, 260 other ranks, 3 women, 1 Greek, 1 Chinaman, 10 distressed merchant seamen, 2 children

and 1 dog! See Beevor, op. cit., note. p216.

5. C.J. 'Jack' Hamson was a member of Pendlebury's initial band and was captured with a detachment from the Argyll and Sutherland Highlanders near Tymbaki. He was particularly scathing about the ratio of senior officers evacuated (see Beevor, op. cit., p218). After the war he returned to the academic life and became Professor of Comparative Law at Trinity College, Cambridge.

6. Young's adjutant, Michael Borwick who, like his CO, had distinguished himself during the retreat, was overcome with emotion at having to give the order for surrender to the men under his command after all had fought so hard and come so far.

7. Anthony Beevor recounts the story of Colonel Walker's meeting with an officer of the 100 Mountain Regiment, an Austrian. 'What are you doing here, Australia?' the officer enquired. 'One might ask what are you doing here Austria?' Walker replied. 'We are all Germans,' he was told.

8. A.H. Whitcombe quoted in MacDonald, op. cit., p293.

9. R. H. Thompson, ibid., p293.

10. ibid., p293.

Chapter 10

1. Quoted in Beevor, op. cit., p336.
2. Quoted in MacDonald, op. cit., p301.
3. ibid., p303.
4. Quoted in Simpson, op. cit., pp278-9.
5. ibid., pp279-80.
6. ibid., p279.
7. Clark, op. cit., p203.

Appendix 1

Order of Battle

1. Allied Forces

Creforce HQ – Major-General B. Freyberg

> C Squadron the King's Own Hussars
> B Squadron the Royal Tank Regiment
> 1st Battalion Royal Welch Fusiliers (Force Reserve)

HQ New Zealand Division ("NZ") – Brigadier E. Puttick

> 27th NZ MG Battalion
> 5th NZ Field Artillery Regiment

4th NZ Infantry Brigade – Brigadier L.M. Inglis:

> 18th NZ Infantry Battalion
> 19th NZ Infantry Battalion
> 20th NZ Infantry Battalion
> 1st Light Troop Royal Royal Artillery ("RA")

5th NZ Infantry Brigade – Brigadier J. Hargest:

> 21st NZ Infantry Battalion
> 22nd NZ Infantry Battalion
> 23rd NZ Infantry Battalion
> 28th (Maori) Infantry Battalion
> 7th NZ Field Company
> 19th Army Troops Company
> 1st Greek Regiment

10th NZ Infantry Brigade – Colonel H.P. Kippenberger:

NZ Divisional Cavalry
NZ Composite Battalion
6th Greek Regiment
8th Greek Regiment

HQ 14th Infantry Brigade – Brigadier B.H. Chappel

2nd Battalion the Leicestershire Regiment
2nd Battalion the York and Lancaster Regiment
2nd Battalion the Black Watch
2/4 Australian Infantry Battalion
1st Battalion Argyll and Sutherland Highlanders
7th Medium Regiment RA (deployed as infantry)
3rd Greek Regiment
7th Greek Regiment
Greek Garrison Battalion

HQ 19th Australian Infantry Brigade – Brigadier G.A. Vasey

2/3 Field Artillery Regiment RAA
2/1 Australian Infantry Battalion
2/11 Australian Infantry Battalion
2/7 Australian Infantry Battalion
2/6 Australian Infantry Battalion
4th Greek Regiment
5th Greek Regiment
Greek Gendarmerie

HQ Mobile Naval Base Defence Organisation – Major-General C.E.
Weston

15th Coastal Defence Regiment RA
Royal Marine Battalion
1st Ranger Battalion (9th Battalion KRRC)
Northumberland Hussars
106th Royal Horse Artillery
16th Australian Brigade Composite Battalion
17th Australian Brigade Composite Battalion
1st 'Royal Perivolian' Composite Battalion
2nd Greek Regiment

2. Axis Forces

HQ Fliegerkorps XI – Major-General K. Student

GGzbV 1,2 and 3 (Ju-52)
22nd Air Assault Division (deployed in Romania)

HQ 7th Flieger Division – Lieutenant-General W. Sussman

(7th) Engineer, Artillery, Machine Gun, Anti-Tank, AA and Medical battalions

1st Parachute Regiment – Major B. Brauer;

1st Battalion – Walther
2nd Battalion – Burkhardt
3rd Battalion – Schulz

2nd Parachute Regiment – Major A. Sturm:

1st Battalion – Kroh
2nd Battalion – Schirmer
3rd Battalion – Weidermann

3rd Parachute Regiment – Major R. Heidrich:

1st Battalion – Heydte
2nd Battalion – Derpa
3rd Battalion – Heilmann

HQ Air Assault Regiment – Major-General E. Meindl

1st Battalion – Koch
2nd Battalion – Stenzler
3rd Battalion – Scherber
4th Battalion – Gericke

HQ 5th Mountain Division – Major-General J. Ringel

(95th) Artillery, Anti-Tank, Reconnaissance, Engineer and Signals Battalions

85th Mountain Regiment – Krakau:

    1st Battalion
    2nd Battalion
    3rd Battalion

100th Mountain Regiment – Utz:

    1st Battalion
    2nd Battalion
    3rd Battalion

141st Mountain Regiment – Jais:

    1st Battalion
    2nd Battalion
    3rd Battalion

Appendix 2

The Commandments of the Parachutist

Each trooper had a memorandum of these sewn into the lining of his pack. They are a mix of Teutonic sentiment and sound advice:

1. You are the chosen ones of the German army. You will seek combat and train yourselves to endure any manner of test. To you the battle shall be fulfilment.
2. Cultivate true comradeship, for by the aid of your comrades, you will conquer or die.
3. Be aware of talking. Be not corruptible. Men act while women chatter. Chatter may bring you to the grave.
4. Be calm and prudent, strong and resolute. Valour and the enthusiasm of an offensive spirit will cause you to prevail in the attack.
5. The most precious thing in the presence of the foe is ammunition. He who shoots uselessly, merely to comfort himself, is a man of straw who merits not the title of parachutist.
6. Never surrender. To you death or victory must be a point of honour.
7. You can triumph only if your weapons are good. See to it that you submit yourself to this law – first my weapon and then myself.
8. You must grasp the full purpose of any enterprise, so that if your leader is killed you yourself can fulfil it.
9. Against an open foe fight with chivalry, but to a guerrilla extend no quarter.
10. Keep your eyes wide open. Tune yourself to the topmost pitch. Be as nimble as a greyhound, as tough as leather, as hard as Krupp steel, and so you shall be the German warrior incarnate.

Appendix 3

The Intelligence War

[Freyberg] had known nothing of Ultra until Wavell appointed him to command in Crete and so he was quite without experience in interpreting it. Yet almost at once he was compelled by events to make operational decisions in the light of it, without the benefit of a second opinion or any advice whatever [Group Captain Beamish, the Ultra intermediary on Crete, was not in the chain of command]. [Moreover] *in the whole course of history no island had ever been captured except from the sea* [author's italics] the only evidence that the new airborne arm could overpower ground defences consisted of [the evidence from Eben Emael and associated minor operations]. The first parachute battalions in the British army would not be formed for another six months. Finally the fact that the Royal Navy's command of the Mediterranean was being seriously challenged for the first time since Nelson's victory over the French in Aboukir Bay in 1798 was in itself enough to reinforce fears of attack by the traditional means... In spite of Ultra [Freyberg's] apprehension of danger from the sea can only be faulted by an abuse of hindsight.[1]

Ultra, Britain's best kept secret in 1941, was born of Enigma and this was the brainchild of a German inventor Arthur Scherbius whose objective had been to design a machine that could both encipher and decipher automatically. The concept was not a novel one, the machines all worked on a rotating disc principle and Scherbius was not the first to attempt a mass produced version.

His design came onto the market in 1923 and was adopted by the German army five years later. Its capabilities were sufficient to also enthuse the Navy and, latterly the Luftwaffe. Its value was for usage in all secret messaging that was vulnerable to interception – primarily radio traffic.

What, in modern business parlance, would rank as the Engima's unique selling point 'USP' was its reflector disc which empowered the machine to both crypt and decrypt. Powered by dry cell batteries, it was also lightweight and easily portable, in appearance as innocuous as a contemporary portable typewriter. It's potential in the field of modern warfare was immense.

Because of its ability to multiply possible encryption's to such an infinite degree it was thought utterly impregnable to 'cracking' by even the most gifted of cryptanalysts, the Germans believed that hundreds or even thousands of mathematicians could labour for generations without success. This belief persisted even once the codes had been thoroughly penetrated.

It was not in fact the 'boffins' at the British Government's Code and Cipher School (GCCS) at Bletchley Park who first began to fracture the Enigma but the cryptanalysts of the Polish army who achieved miracles by means, mainly of pure applied mathematics with some 'mechanical aids'. The latter, which were to assume increasing importance at Bletchley, were electro mechanical devices that tested the solutions of encrypts much faster than could be achieved by pure manpower – 'bombes' as they were known.

In July 1939 as the threat of war loomed British and French Intelligence officers attended their Polish counterparts in Warsaw where they were presented with a facsimile of the Engima the ingenious Poles had constructed. However, they'd been beaten by the Enigma designers who'd now added two additional discs thus multiplying the complexity.

Room 40 at Bletchley continued on the Polish model with the recruitment of civilian academics, an eccentric and eclectic mix of genius whose existence became the stuff of legend. The most outstanding of these was Alan Turing whose Olympian intellect stood out even in such gifted company. He could be described as the originator of the computer and it was he who designed the bombes at Bletchley.

An element of operator laziness and, in the case of Luftwaffe personnel, inexperience, greatly facilitated the cryptanalysts work as did the use of some partial guessing or 'cribs' as they were known. The intellectual powerhouse that was Hut 40, backed by thousands of hours of work, achieved miracles, firstly the Luftwaffe codes were broken and then others.

Some of the naval codes and that used by the Gestapo were

never cracked. 'Shark' the Atlantic U Boat key remained inviolate all through the murderous months of 1941 - 1942 when the Battle of the Atlantic hung in the balance and thousands of seaman with hundreds of thousands of tons of merchant shipping succumbed to U Boat attacks.

One of the limitations with ULTRA as the Enigma intelligence was designated was that officers in the field never saw the original decrypts, these were frequently unintelligible and needed to be translated and, effectively, interpreted, so that which was passed on was in an edited form. In the early days this fine art of interpretation was left to linguists rather than intelligence officers (IOs) with the risk that a competent IO might have read the decrypt in a subtly different way.

As ULTRA was Britain's most closely guarded secret, perceived as the trump that could tip the finely wrought scales between survival and defeat, entry to the circle of initiates was restricted. The source of the intelligence had to be concealed so that the Germans would not come to suspect Enigma had been broken. Field commanders were generally told that information had been gleaned from well placed agents or detailed reconnaissance.

Freyberg was not within this charmed circle, ULTRA intelligence was thus filtered through Cairo and, as early as 1 May there was an intimation of the intended attack on Crete, four days later further intercepts revealed the main targets and set a date the 17th when the assault would begin. This signals traffic, crucially, mentioned seaborne elements and made reference to 5th Mountain Division.

Creforce HQ had been set up in a disused quarry above Souda Bay, utilising a network of caves that offered good protection from aerial bombardment. Freyberg's staff was somewhat makeshift, with a chronic shortage of signallers and reliable wireless sets. Weston, huffed at his removal from overall command, retained a separate and well equipped HQ. Freyberg was too punctilious to 'pull rank'. In the troglodyte world of Creforce HQ Captain Sandover was the IO responsible for decrypting the ULTRA intercepts.

It must, therefore, be borne in mind that Freyberg was not in the 'know' where this marvellous intelligence was coming from, the signals were codenamed 'OL' for 'Orange Leonard' the usual fiction about agents in place being employed. The key Engima decrypt, (OL 2/302), was passed to Creforce at 5.45 p.m. on 13

May and, on the surface, was pure gold.

This confirmed that the date of the attack was to be the 17th as previously understood, (this was moved to the 20th subsequently), it specified the first day's targets for the paratroops as Maleme, Chania, Rethymnon and Heraklion. It revealed the extent of the air support that would be thrown into the fight, that additional troops would be brought in by glider and, latterly, once an airstrip was secured, by transports. It finally confirmed that elements of the projected invasion force would arrive by sea, together with AA batteries.

In addition 12th Army will allot three Mountain Regiments as instructed. Further elements consisting of motor-cyclists, armoured units, anti-aircraft guns will also be allotted ... Transport aircraft of which a sufficient number – about 600 – will be allotted for this operation, will be assembled on aerodromes in the Athens area. The first sortie will probably carry parachute troops only. Further sorties will be concerned with the transport of the air landing contingent, equipment and supplies, and will probably include aircraft towing gliders ... the invading force will consist of some 35,000 men, of which some 12,000 will be in the parachute landing contingent and 10,000 will be transported by sea ...Orders have been issued that Souda Bay is not to be mined, nor will Cretan aerodromes be destroyed, so as not to interfere with the operation intended.[2]

Although the units to be employed in the assault were listed there was no specific mention of which units would appear where or how, precisely, they were to be landed. ULTRA, despite the very precision of the intelligence actually misinformed Freyberg or allowed him to form a wrong assessment – that a substantial element of the attacking force would be amphibious.

Given that Crete was an island and that, as mentioned, the concept of vertical envelopment was untried, this was not an unreasonable conclusion. Had the fine detail shown that only relatively minor elements of the force would be coming in ships then the General might have re-considered his defensive strategy. The decisions made during the critical period of 21/22 May need to be considered in the light of this. Freyberg was convinced that the initial airborne landings were merely an overture and that the more solid threat would come over the water.

To understand the apparent lethargy that seems to have

informed decisions at 5 Brigade HQ during the crucial battle of 20 May we need to understand that everyone believed that strong coastal defences were required and that the air drop was but the first phase in a combined air and sea operation. Given the intelligence that had been provided and the natural assumptions surrounding the defence of an island fortress, it is difficult to overly criticise the decisions taken, one of which, significantly was that there were insufficient troops available to establish a viable presence west of the Tavronitis – a key failure.

If we consider Ralph Bennett's appraisal of Freyberg's position it is difficult not to agree, only with the benefit of hindsight can we appreciate the seriousness of the error. Given the knowledge he possessed at the time, prior to the attack, Freyberg would have appeared grossly negligent had he not given considerable thought to coastal defence. As John Keegan also points out:

Bennett's reflections on the early work of Hut 3 – precisely in the Crete period – are of the greatest relevance to the understanding of Freyberg's conduct of the battle, since they disclose the serious shortcomings of the ULTRA messages he was sent. What these messages reveal appears to be a complete picture of the impending airborne invasion. What they do not disclose, crucially, is who is going to land where. The objectives – Maleme, Rethymno, Heraklion are given; so is the strength of the force, 7th Parachute Division, a reinforced 5th Mountain Division. The units of the force are not, however, matched with the target zones. The crucial synthesis of the German operation order OL 2/302 of 13 May 1941, the work of Bletchley interpreters, not the transcript of the German intercepts themselves, leaves it unspecified how the Assault regiment and the nine battalions of the Parachute Regiments, 1, 2 and 3 are to be allotted between targets.[3]

Notes
1. Ralph Bennett, the authority on Ultra, (and himself a veteran of Bletchley), as quoted in Keegan, pp195-6.
2. Quoted in Keegan, pp193-4.
3. Keegan, pp207-8.

Appendix 4

The King of the Hellenes

King George of Greece, by the end of April 1941, was a king largely without a kingdom. He was evacuated from Athens, together with his immediate entourage, by flying boat. His arrival did not spark any great enthusiasm amongst the pro-republican Cretans. The common sentiment was revulsion at the man who had acquiesced to the Metaxas coup in 1936 and, even now, brought with him the hated Maniadakis, Minister for Security, essentially a fascist thug who was bolstered by a full platoon of his equally thuggish secret police.

As a sop to the islanders' sensibilities, Emmanuel Tsouderas, a native Cretan, politician and financier, was appointed as Prime Minister in the wake of Koryzis' suicide. Most, however, saw him as little better than a turncoat, a poor substitute for the whole of the Cretan 5th Division whose heroism had been rewarded with abandonment. General Papastergiou, the divisional commander was, however, evacuated.

For this betrayal the locals ensured he was duly assassinated the moment he set foot on the streets of Chania. British diplomats regarded this sudden taking off of the despised and discredited general as a healthy sign! Cretan morale was not affected by the loss of the mainland.

As his official residence the King chose the Villa Ariadne, Evans' elegant Edwardian villa by Knossos, built once the present ruler's uncle, Prince George, ceded the freehold of the site in 1900. Pendlebury had been curator from 1930-1934. Here he established his truncated court, joined by Princess Katherine and other members of the Royal Family, (most of whom transferred to Egypt by flying boat prior to the attack). Though ousted from the bulk of his domain King George had ensured that the country's gold reserves travelled with him.

Congenial as Villa Ariadne might be it was too far removed from the hub of civic and diplomatic life in Chania whence the court was presently removed. British observers pointed out that the presence of the brutal Maniadakis was an affront to local, republican sympathies and he, together with his henchmen, was shifted to Egypt where he was able to continue terrorising the pro-Venizelist Greek community.

In a further, doomed attempt to win hearts and minds the King appointed two Cretan officers, Generals Zannakis and Skoulas as Minister of War and commander of local forces respectively. As, however, the Greek troops on the island had all been placed under Freyberg's own hand these sinecures held little currency.

As far as the Allied commander-in-chief was concerned the presence of the King within his enceinte was an embarrassment, yet another millstone of responsibility for which he could perceive little or no point. Freyberg would have preferred the King, his family and hangers on removed to the safety of Cairo but neither Wavell nor indeed Churchill would hear of it. King George was seen, somewhat wishfully, as a talisman, a guarantee of Allied legitimacy and, optimistically, a symbol to his people.

In reality the King was a symbol of oppression, adrift in a sea of republicanism, his tame surrender to Metaxas and the anti-republican measures which ensued had earned him nothing but opprobrium.

The Cretans particularly felt the confiscation of their firearms as a form of ritual emasculation. To the proud *Palikari* the gun was a symbol of independence, identity, liberty, even of manhood; to rob them of their arms was to forfeit any hope of allegiance. When, after the end of the war, in 1946, a plebiscite on the role of monarchy was held, the islanders overwhelmingly voted against. Equally they would have no truck with the communists, preparing to rend their battered country further with bloody civil strife.

In the halcyon interlude prior to 20 May the King spent his days touring his shrunken domain, his large limousine, complete with pennons, a prime target of opportunity for prowling Messerschmitts, adding a Ruritanian note to the drab attire of war. As the bombing intensified the court was moved from the dangerous environs of the beleaguered island capital to the supposedly safer isolation of Perivolia, near Galatos. Here the King was guarded by 12 Platoon, B Company, 18th Battalion, commanded by Lieutenant W.H. Ryan.

The events of the morning of 20 May shattered this meaningless if not unpleasant round. Ryan ushered his charge out of the villa and up into the hills behind as the sky filled with parachutists. None too soon; by the end of the day the area was in German hands, and the party was in full flight toward the central massif of the White Mountains. It is unlikely the invaders had any immediate notion of how valuable a prize had slipped from their clutches but the party faced a gruelling trek over the high snow bound plateau atop the Lefka Ori before the difficult descent to the south coast and evacuation aboard HMS *Decoy*.

On the 21st they struggled for fourteen long hours to gain the summit, the sounds and sights of battle spread over the coastal plain below; the King, Prince Peter, M. Tsouderas, the President of the National Bank and a random rump of the attendant entourage bolstered by 12 Platoon.

Throughout this difficult and potentially dangerous journey, certainly the greatest test of any personality, the King remained a tower of optimism. He chatted and joked with the Australians, not likely to be easily impressed by a title, yet who clearly came to respect and like their charge who shared their hardships, the toil of the barren march, the meagre rations with constant good humour.

Despite the constant, harassing presence of German observation aircraft, the ubiquitous Fiesler Storch, the whole party eventually completed the difficult descent to the south coast without lost and were duly taken off by sea.

Bibliography

Antill, P.D., Crete 1941, Osprey 'Campaign' Series England, 2005

Barker, A.J., *British and American Infantry Weapons of World War Two*, London, 1969

Barker, A.J., *German Infantry Weapons of World War Two*, London, 1972

Beevor, A., *Crete: The Battle and the Resistance*, London, 1991

Chant, C., *Airborne Invasions*, England, 1976

Clark, A., *The Fall of Crete*, London, 1962

Clark, A., *Barbarossa*, London, 1965

Davis, B.L., *German Army Uniforms and Insignia 1939 – 1945*, London, 1971

Durrell, L., *The Greek Islands*, London, 1978

Garnett, D., *The Campaign in Greece and Crete*, HMSO, 1942

Gordon-Douglas, S.R., *German Combat Uniforms 1939 – 1945*, Almark 'Uniform' Series, London, 1970

Keegan, J., *Intelligence in War*, London, 2003

Kershaw, A., *Weapons of War*, Purnell's History of the Second World War, 1973

Kershaw, A. (ed.), *The Tank Story*, Purnell, 1972

King, J.B. and John Batchelor, *Infantry at War 1939 –1945*, Purnell, 1973

MacDonald, C., *The Lost Battle; Crete 1941*, London, 1993

Moss, W. Stanley, *Ill Met by Moonlight*, London, 1999.

Psychoundakis, G., *The Cretan Runner*, London, 1998

Simpson, A., *Operation Mercury – the Battle for Crete 1941*, London, 1981

Index

Stackpole Military History Series

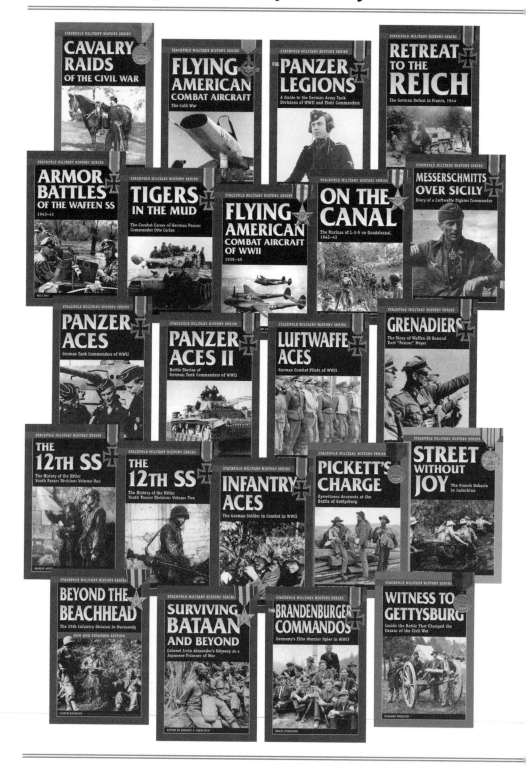

Real battles. Real soldiers. Real stories.

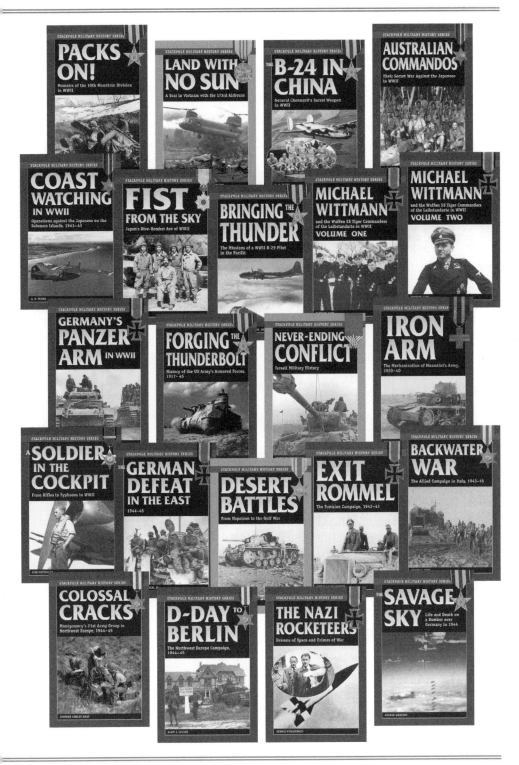

Stackpole Military History Series

Real battles. Real soldiers. Real stories.

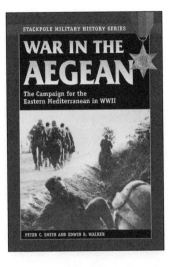

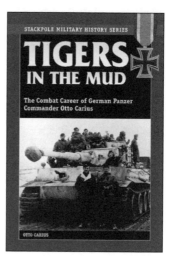

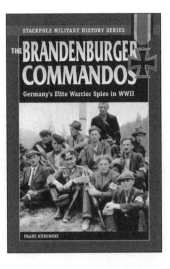

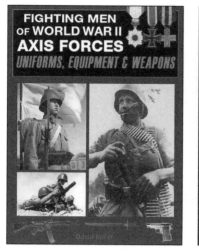 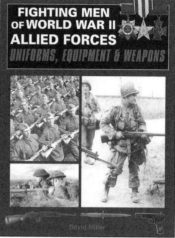